7th International Conference on Nanomaterials and Materials Engineering (ICNME 2019)

IOP Conference Series: Materials Science and Engineering
Volume 761

Chengdu, China
25 - 27 October 2019

ISBN: 978-1-7138-2662-0
ISSN: 1757-8981

Printed from e-media with permission by:

Curran Associates, Inc.
57 Morehouse Lane
Red Hook, NY 12571

Some format issues inherent in the e-media version may also appear in this print version.

Institute of Physics
Dirac House, Temple Back
Bristol BS1 6BE UK

Phone: 44 1 17 929 7481
Fax: 44 1 17 920 0979

techtracking@iop.org

Additional copies of this publication are available from:

Curran Associates, Inc.
57 Morehouse Lane
Red Hook, NY 12571 USA
Phone: 845-758-0400
Fax: 845-758-2633
Email: curran@proceedings.com
Web: www.proceedings.com

TABLE OF CONTENTS

BATTERY MATERIALS

BIOLOGICAL MATERIALS

NANOMETER MATERIALS

MATERIAL PHYSICS

IOP Conf. Series: Materials Science and Engineering **761** (2020) 011001 doi:10.1088/1757-899X/761/1/011001

Welcome Message from Conference Chair

ICNME 2019

Materials are the basis for human survival and development. New materials such as metal materials, nanomaterials, energy materials, magnetic materials, polymer materials and biomedical materials are the cornerstone and pioneer of the development of high-tech industries. They play an increasingly important role in the development of the national economy. Thanks to the special properties of nanomaterials, scientists have made breakthroughs in various fields of study, which has also contributed to the increasing widespread application of nanomaterials. Nanomaterials are widely applied in the production of special performance materials, biomedical aspects, nano-biocomputer development, as well as in sea water purification, aerospace, environmental energy, microelectronics, etc.

7th International Conference on Nanomaterials and Materials Engineering was held successfully in Sichuan University at Chengdu, China during October 25-27, 2019. It is mainly organised by International Academy of Computing Technology (IACT) and hosted by Sichuan University and Panzhihua University. The proceedings of ICNME2019 includes accepted papers which have been submitted to the conference from universities, research institutes and industries. The acceptance of these papers is based on their quality and their relevance to the conference. The proceedings disseminate, to readers, the latest advances in the broad fields of Nanomaterials and Nanotechnology, Biomaterials and Biomedical Engineering, Ceramic and Glass Materials, Computational Material Science, Coupling of Multiple Ferroic Order Parameters, etc.

During the keynote session, Prof. Yanwei Ma from Institute of Electrical Engineering, Chinese Academy of Sciences, China shared with us his talk on *High-Performance Graphene-based Supercapacitors: From Materials Design to Devices Fabrication;* Prof. Ying-Hao Eddie Chu from National Chiao Tung University, Taiwan gave a speech on *Van der Waals Oxide Heteroepitaxy for Transparent Flexible Electronics;* Prof. Dewei Zhao, from Institute of Solar Energy Materials and Devices, College of Materials Science and Engineering, Sichuan University, China made a speech on *Efficient Perovskite-Perovskite Tandem Solar Cells.* For the parallel session, 3 invited speakers also offer the invited speech, Dr. Ruijuan Xiao from Institute of Physics, Chinese Academy of Sciences talked on *The acceleration of discovery new fast ionic conductors by high- throughput calculations;* Prof. Liang Qiao from University of Electronic Science and Technology of China, China talked on *Octahedral Engineering and Interfacial Structure of Heteroepitaxial Complex Oxides;* Dr. Deyang Chen from South China Normal University, China talked on *Strain engineering of epitaxial oxide heterostructures beyond substrate limitations.*

IOP Conf. Series: Materials Science and Engineering **761** (2020) 011001 doi:10.1088/1757-899X/761/1/011001

Concurrent sessions and a poster session cover a wide range of topics and issues, including both contributed papers, all with a central focus of management. Topics include the Material Mechanics and Model; Engineering Materials and Properties; and Material Properties and Preparation Methods etc.

We take this opportunity to thank all committee members who have devoted their time, effort and wonderful works to the technical evaluation of the received papers as well as the organization of the conference. At the same time, we would like to thank all the attendees, for their effort of coming to Chengdu to attend the conference during their busy schedules, and to share their research findings and knowledges during the conference sessions. We hope that everyone could find ICNME2019 to be intellectually stimulating.

Prof. Xiaohong Zhu, Sichuan University, China

Conference Chair of ICNME 2019

IOP Conf. Series: Materials Science and Engineering **761** (2020) 011002 doi:10.1088/1757-899X/761/1/011002

Conference Committees

International Advisory Committees
Prof. Mohamed Henini, University of Nottingham, UK
Prof. Galina Ilieva, University of Plovdiv, Bulgaria

Conference Chair
Prof. Xiaohong Zhu, Sichuan University, China

Conference Program Chairs
Prof. Ying-Hao Eddie Chu, National Chiao Tung University, Taiwan
Prof. Guneik Jang, Chungbuk National University, Korea
Prof. Tjokorda Gde Tirta Nindhia, Udayana University, Indonesia

Conference Local Chair
Prof. Jiliang Zhu, Sichuan University, China

Conference Publicity Chairs
Prof. Seong Soo Choi, SunMoon University, Korea
Assoc. Prof. Brian Yuliarto, Institut Teknologi Bandung, Indonesia

Conference Technical Committees
Prof. Anders Gåård, Karlstad University, Sweden
Prof. Aidy Ali, National Defence University of Malaysia, Malaysia
Prof. Geetesh Goga, K.C. College of Engg. & I.T., India
Prof. Dr. Osman ADIGUZEL, Firat University, Turkey
Prof. Sroisiri Thaweboon, Mahidol University, Thailand
Assoc. Prof. Mohd Rozi Ahmad, Universiti Teknologi MARA (UiTM Shah Alam), Malaysia
Assoc. Prof. S. Balasivanandha Prabu, Anna University, India
Assoc. Prof. Abbas Teimouri, Payame Noor University (PNU), Iran
Asst. Prof. Dr. Rafia Usman Khan, NED University, Pakistan
Assoc. Prof. Tianwei Li, Shenyang Jianzhu University, China
Dr. T. THEIVASANTHI, Kalasalingam University, India
Dr. Syed Bahari Ramadzan Syed Adnan, University of Malaya, Malaysia
Dr. Nirali Gondaliya, SVM institute of Technology, India
Dr. Muhammad Irfan-ul-Hassan, University of Engineering and Technology, Pakistan
Dr. Abid Ali Shah, International Islamic University, Pakistan
Dr. Arjon Turnip, Indonesian Institute of Sciences, Indonesia
Dr. Azmah Hanim Mohamed Ariff, Universiti Putra Malaysia, Malaysia
Dr. Shanmuga Priya Natesan, Siddaganga Institute of Technology, India
Dr. Ping Xiang, City University of Hong Kong, Hong Kong

IOP Conf. Series: Materials Science and Engineering **761** (2020) 011002 doi:10.1088/1757-899X/761/1/011002

First name	Surname	Institution	E-mail address
Guangfu	Liu	Northeastern University, China	gfliu_neu@126.com
Jun	Wang	Chongqing University of Science and Technology, China	wangjunpaper@126.com
Sroisiri	Thaweboon	Mahidol University, Thailand	sroisiri.tha@mahidol.ac.th
Tianwei	Li	Shenyang Jianzhu University, China	fighting_5@163.com
Jieru	Wang	Zhejiang University, China	1152657977@qq.com
Tjokorda Gde Tirta	Nindhia	Engineering Faculty Udayana University, Indonesia	nindhia@yahoo.com
Yunchao	Wang	Jimei University, China	ychaowang@jmu.edu.cn
Yi Jing	Gu	Guizhou University, China	yijing-gu@163.com
IAI	Qi	Panzhihua University, China	49129834@qq.com
Xueping	Luo	Panzhihua University, China	1044145023@qq.com
Chuanhai	Xiao	Panzhihua University, China	386277465@qq.com
Mohand	Berdja	Unité de Développement des Equipements Solaires (UDES), Centre de Développement des Energies Renouvelables (CDER), R.N n°11, BP 386, 42415 Bou Ismail, Tipaza, Algeria.	berdjamohand@gmail.com
Lei	Liu	Panzhihua University, China	superware@163.com
Liang	Zou	Shandong University, China	zouliang@sdu.edu.cn
Tao	Chen	Shanghai University, China	chentao0709@shu.edu.cn

IOP Conf. Series: Materials Science and Engineering **761** (2020) 011004 doi:10.1088/1757-899X/761/1/011004

Peer review statement

All papers published in this volume of *IOP Conference Series: Materials Science and Engineering* have been peer reviewed through processes administered by the proceedings Editors. Reviews were conducted by expert referees to the professional and scientific standards expected of a proceedings journal published by IOP Publishing.

IOP Conf. Series: Materials Science and Engineering **761** (2020) 012001 doi:10.1088/1757-899X/761/1/012001

Synthesis and Electrochemical Performance of Nano-sized CuO/TiO2 for Lithium-ion Batteries

Guangfu LIU, Kuiren LIU, Zhongbao SHAO, Xuetian LI
Deparment of Nonferrous Metallurgy, School of Metallurgy, Northeastern University, 3 Wenhua Road, Shenyang, 110819, China

gfliu_neu@126.com

Abstract. CuO/TiO2 nanopowder was successfully synthesised by precipitation method. XRD, SEM and galvanostatic testing were applied to study the crystal structure and electrochemical performance of CuO/TiO2. The results showed that the prepared sample was highly-crystalline and exhibited high capacities and well cycle stabilities. The initial discharge capacity of CuO/TiO2 reached 566.5 mAh/g, and retained at 375.9 mAh/g after 20 cycles at 0.1C-rate. It is implied that CuO/TiO2 nanopowder prepared by precipitation method coulde be used as an anode material for lithium-ion batteries.

1. Introduction

Lithium-ion battery (LIB) is undoubtedly the fastest-growing power source available in various fields (traffic power supply, power storage, and mobile communication power supply) today [1]. Since the rapid commercial demands for low cost, green, safety, and more powerful batteries, many scientists have devoted to improve the performance of conventional anode materials [2-4].

Recently, transition-metal oxides have been widely studied as anode material for lithium-ion batteries for decades owing to their special properties [5]. Among varies of tansition-metal oxides applied for anode materials in lithium-ion batteries, Copper oxide (CuO) has drawed much attention due to its high capacity (674mAh/g), chemical safety, low cost, environment benignity and abundance in nature [6]. However, CuO which react with Li would bing large volume variation during Li-alloying/dealloying process, and result in short cycle life. This shortage largely hinder his practical applications [7, 8]. To sovle this problem, a variety of methods have been explored, including preparing nano-scale materials, controlling structures, and combining with other complex conductive materials [9-12].

As for composite materials, TiO2 has proven to be a highly promising candidate due to its small degree of volume change (<4%) during the process of Li-insertion/extraction [13-15]. Besides, TiO2 also has a good performance in capacity (350mAh/g) which was a potentially beneficial to both the structural stability and the electrochemical performance of composites materials [16].

To achieve this excellent cycling performance of anode materials, we planned the synthesis of a CuO/TiO2 nanopowder using precipitation method. We expected that CuO/TiO2 nanopowder will improve the electrochemical performance as an anode material in lithium-ion batteries.

2. Experiments

2.1. Samples preparation

IOP Conf. Series: Materials Science and Engineering **761** (2020) 012001 doi:10.1088/1757-899X/761/1/012001

First, 48.3 g of copper (II) nitrate trihydrate (Cu (NO3)2·H2O, AR, 99.0%) and 4.8 g of titanium (IV) sulfate (Ti (SO4)2, AR, 97.0%) were dissolved in 100 mL deionized water under ultra-sonic dispersion to be completely dissolved for 30 min. Then Ammonium bicarbonate (NH4HCO3, AR, 99.9%) (1.0 mol/L) as a precipitant was added into the solution above to prepare the precursor of Cu and Ti, and stirred with a speed of 300 rpm at room temperature, and stopped dropping the precipitation when the pH of solution reached to 9, and then separated by filtration and washed by deionized water for several times until neutralization.The filter we got was dried in an oven at 105 °C for 12 h. Finally, CuO/TiO2 nanopowder was obtained after treating thermally in a furnace at 500 °C for 3h. The synthesis process of CuO/TiO2 nanopowder is shown in Figure 1.

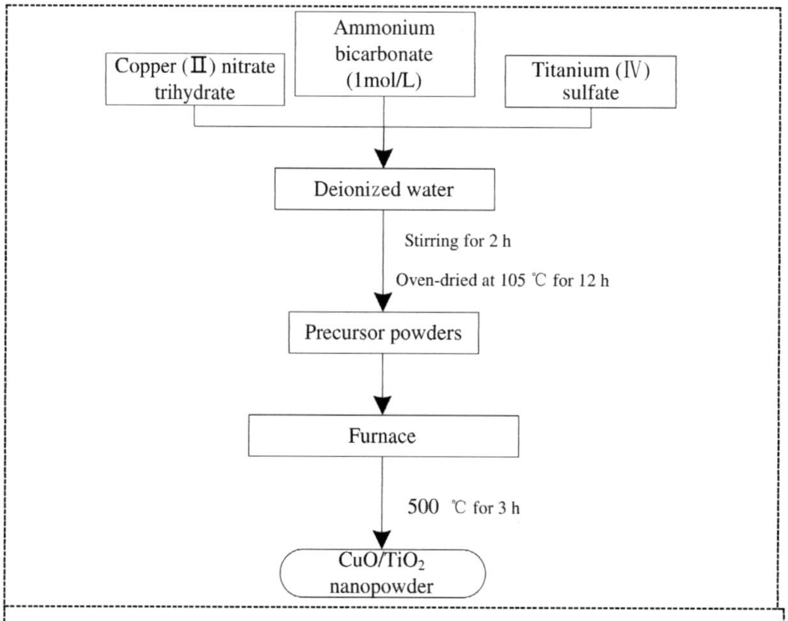

Figure 1. Flow sheet of the process route.

2.2. Characterization

The structure of CuO/TiO2 sample was obtained by X-ray diffraction (XRD, Rigaku D/max-RB). The morphologie of CuO/TiO2 sample was characterized by scanning electron microscopy (SEM, FEI Nova NanoSEM 430).

2.3. Electrochemical Characterization

The prepared material, carbon black and polyvinylidene fluoride (PVDF) binder (80 wt.%:10 wt.%:10 wt.%) dissolved in N-methyl pyrrolidone (NMP) as electrode slurry was coated on Cu foil (the weight of active material was about 3.5-4.0 mg) heated at 80 °C under vacuum for 12 h. 1 M LiPF6/EC+DEC+EMC (1:1:1 in volume), Celgard 2400 screens and lithium metal foils were used as the electrolyte, membrane and anode, respectively.

The battery-testing system (Neware, China) was applied to test cycle charge-discharge for button battery from 0.01 to 3 V (1C equals to 670 mAh/g).

3. Results and discussion

Figure 2 shows the XRD patterns of CuO/TiO2 sample. It can be comfirmed that after precipitation teaction and thermal oxidative annealing, peaks of tenorite CuO (JCPDS No. 48-1548) and the peaks of anatase TiO2 (JCPDS No. 21-1272) were found in the XRD patterns, which were strong and clean, indicating that the CuO and TiO2 nanostructure was pure and well crystallized without any appreciable impurity after annealing under air conditions.

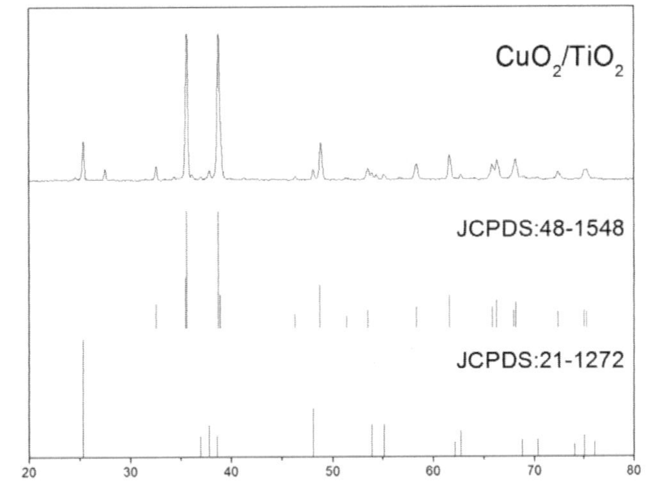

Figure 2. XRD patterns of CuO/TiO_2 and two standard cards (tenorite CuO and anatase TiO_2).

Figure 3 shows the SEM image of CuO/TiO2 sample. The dispersion of CuO/TiO2 powder with sizes of around 50-100 nmand with an irregular shape could be obviously found. The small particle size, which could increase the diffusion rate of Li^+.

Figure 3. SEM image of CuO/TiO_2.

Table 1. Discharge capacities of 1st, 2nd, 5th and 20t

discharge times	discharge capacity
1st	566.5
2nd	476.3
5th	388.1
20th	375.9

Figure 4 and table 1 show the cycling performance of sample CuO/TiO2 electrode. It was apparent that the initial discharge capacity of CuO/TiO2 sample was 566.5 mAh/g, with capacity retention of 66.4% at 0.1C-rate over 20 cycles.

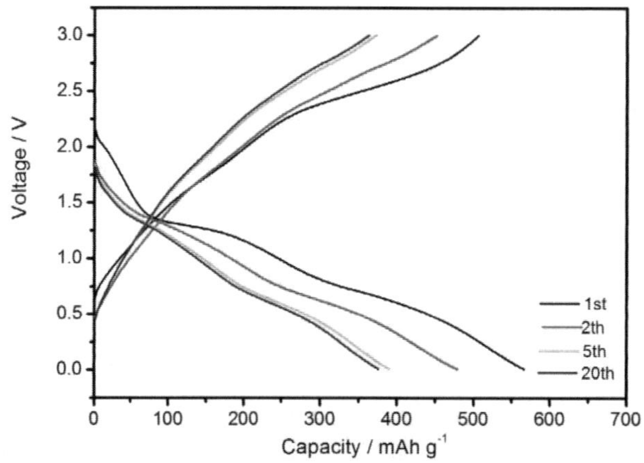

Figure 4. Charge and discharge curves of CuO/TiO_2 for the 1st, 2nd, 5th and20th at 0.1C-rate

4. Conclusion

In conclusion, we prepared CuO/TiO2 nanopowder via precipitation method, and the CuO/TiO2 nanopowder was well complexed. The cycling performance of sample CuO/TiO2 electrode showed that it's an effective way to enchance the the electrochemical properties of CuO, and it provided a novol way to synthesis anode material for lithium-ion batteries.

5. References

[1] Wu Songhao, Fu Gaoliang, Lv Weiqiang, Wei Jiake, Chen Wenjin, Yi Huqiang, Gu Meng, Bai Xuedong, Zhu Liang, Tan Chao, Liang Yachun, Zhu Gaolong, He Jiarui, Wang Xinqiang, Zhang Kelvin H L, Xiong Jie and He Weidong, 2017, "A Single-Step Hydrothermal Route to 3D Hierarchical Cu_2O/CuO/rGO Nanosheets as High-Performance Anode of Lithium-Ion Batteries", *Small*, **14,** 1702667.

[2] Juan Yang, Hangyu Tian, Jingjing Tang, Tao Bai, Lihua Xi, Sanmei Chen and Xiangyang Zhou, 2017, "Self-assembled NiCo2O4-anchored reduced graphene oxide nanoplates as high performance anode materials for lithium ion batteries", *Appl Surf Sci,* **426,**1055-62.

[3] Hai-Jun Peng, Gui-Xia Hao, Zhao-Hua Chu, Cai-Lian He, Xiao-Ming Lin and Yue-Peng Cai, 2017, "Mesoporous spindle-like hollow CuO/C fabricated from a Cu-based metal-organic framework as anodes for high-performance lithium storage", *J Alloy Compd,* **727** 1020-26.

[4] Dae Sik Kim, Dong Jae Chung, Juhye Bae, Goojin Jeong and Hansu Kim, 2017, "Surface engineering of graphite anode material with black TiO_{2-x} for fast chargeable lithium ion battery", *Electrochimica Acta*, **258**, 336-42.

[5] Shuzhen Yang, Yanfang Huang, Guihong Han, Jiongtian Liu and Yijun Cao, 2017, "Synthesis and electrochemical performance of double shell $SnO_{2@}$ amorphous TiO_2 spheres for lithium ion battery application", *Powder Technology,* **322,** 84-91.

[6] Qiming He, Sui Gu, Tian Wu, Sanpei Zhang, Xin Ao, Jianhua Yang, Zhaoyin Wen, 2017 "Self-supported mesoporous FeCo2O4 nanosheets as high capacity anode material for sodium-ion battery", *Chemical Engineering Journal,* **330,**764-73.

[7] Wei Yuan, Zhiqiang Qiu, Yu Chen, Bote Zhao, Meilin Liu, Yong Tang, 2018 "A binder-free composite anode composed of CuO nanosheets and multi-wall carbon nanotubes for high-performance lithium-ion batteries", *Electrochimica Acta,* **267**,150-60.
[8] Shenglian Luo, Fang Sua, Chengbin Liua, Juanxiu Li, Ronghua Liua, Yan Xiao, Yue Li a, Xuanneng Liua and Qingyun Caia, 2011, "A new method for fabricating a CuO/TiO_2 nanotube arrays electrode and its application as a sensitive nonenzymatic glucose sensor", *Talanta,* **86**,157-63.
[9] Chao Chen, Sang Ha Lee, Misuk Cho and Youngkwan Lee, 2015, "Core-shell $CuO@TiO_2$ nanorods as a highly stable anode material for lithium-ion batteries", *Materials Letters,* **140**,111-14.
[10] Martin Leimbach, Christoph Tschaar, Udo Schmidt and Andreas Bund, 2018, "Electrochemical characterization of chromium deposition from trivalent solutions for decorative applications by EQCM and nearsurface pH measurements", *Electrochimica Acta,* **270**,204-09.
[11] Jiansheng Chen, Lin Xu, Ruiqing Xing, Jian Song, Hongwei Song, Dali Liu and Ji Zhou, 2012, "Electrospun three-dimensional porous CuO/TiO2 hierarchical nanocomposites electrode for nonenzymatic glucose biosensing", *Electrochemistry Communications,* **20**,75-78.
[12] Ang Li, Renyue He, Zhuo Bian, Huaihe Song, Xiaohong Chen and Jisheng Zhou, 2018, "Enhanced lithium storage performance of hierarchical CuO nanomaterials with surface fractal characteristics", *Applied Surface Science,* **443**,382-88.
[13] M. Saravanan, Shantikumar V. Nair and Alok Kumar Rai, 2017, "Low temperature synthesis of carbon-wrapped CuO synthesized without using a conventional carbon source for Li ion battery application", *Physica E,* **94**,113-17.
[14] Guangxia Wang, Yongming Sui, Meina Zhang, Man Xu, Qingxin Zeng, Chuang Liu, Xinmei Liu, Fei Du and Bo Zou, 2017, "One-pot synthesis of uniform Cu_2O–CuO–TiO_2 hollow nanocages with highly stable lithium storage properties", *J. Mater. Chem. A,* **5**,18577.
[15] Peng Liu, Xifeng Xia, Wu Lei and Qingli Hao, 2017, "Rational synthesis of highly uniform hollow core–shell $Mn_3O_4/CuO@TiO_2$ submicroboxes for enhanced lithium storage performance", *Chemical Engineering Journal,* **316**,214-24.
[16] *Yong Wang, Dongxia Wang, Qingyuan Li, Wenbin Guo, Fanchao Zhang, Yang Yu, and Yiqing Yang*, 2018, "General Synthesis and Lithium Storage Properties of Metal Oxides/MnO_2 Hierarchical Hollow Hybrid Spheres", *PART PART SYST CHAR,* **35**, 1700336.

Acknowledgments
This research was supported by my supervisor K.R LIU who provided experiment feasibility guidance and experiment facilities during the research.

Three Dimensional Carbon Supported Ultrasmall Fe_2O_3 Nanoparticles as Anode Material for Lithium Ion Batteries

Wang Jieru[1], Xu Junming[2], Zhang Linjuan[2]
[1] School of Materials Science & Engineering, Zhejiang University, Hangzhou 310027, PR China
[2] College of Electronic Information, Hangzhou Dianzi University, Hangzhou 310018, PR China

Abstract. In this article, carbon nanotubes were grown on free standing multi-layer graphene sheet to form three dimensional carbon (3D-carbon) materials, and then used as support of Fe_2O_3 nanoparticles. Ultrasmall Fe_2O_3 nanoparticles were uniformly deposited on the surface of the 3D-carbon. Due to the synergistic effect of Fe_2O_3 nanoparticles and 3D-carbon, the electrochemical properties are higher than that of bare Fe_2O_3 material and multi-layer graphene sheet supported Fe_2O_3. Reversible capacity 528 mAh g-^{1} is maintained at a high current density of 500 mA g^{-1} after 50 cycles. This composite shows a stable cycling and high rate performance.

Keywords. Fe_2O_3; 3D-carbon material; Hydrothermal; Lithium ion battery.

1. Introduction

In order to meet the increasing demand of mobile power, researches on the electrode materials of lithium ion batteries (LIBs) for high energy and power density has become a hotspot. As a typical transition metal oxide (TMO), Fe_2O_3 has been considered as a promising choice for anode material due to its high theoretical specific capacity (~1000 mAh g^{-1}), nontoxicity, abundance and low cost. However, similar to other TMO, they undergo the drawbacks such as poor capacity retention and deprived rate capability. The poor cyclability is caused by a massive volume change in the metal oxides, which leads to electrode swelling/contraction and thus loss of electronic conduction pathways.

Compositing Fe_2O_3 with carbon materials is considered to one effective way to improve the electrical conductivity. Meng *et al* [1] reported the 3D α-Fe_2O_3@graphene aerogel composites. The reversible capacity was maintained at 745 mAh g^{-1} while cycling 100 times at 100 mA g^{-1}. Cheng *et al* [2] successfully synthesize α-Fe_2O_3/C composites and exhibited excellent cycling performance. A reversible capacity of 688 mAh g^{-1} after 50 cycles at 0.2 C was obtained. Wu *et al* [3]fabricated Fe_2O_3/CNTs composites by hydrothermal method and heat treatment, which provided initial reversible capacity of 847.5 mAh g^{-1} at a current density of 100 mA g^{-1} and exhibited a reversible capacity of 865.9

IOP Conf. Series: Materials Science and Engineering **761** (2020) 012002 doi:10.1088/1757-899X/761/1/012002

mAh g^{-1} after 50 cycles. Up to now, the uses of graphene and carbon nanotube have improved the Fe_2O_3 cycling and rate performance as LIB.

Here, we synthesized a kind of 3D-carbon material that CNTs were homogeneously grown on multi-layer graphene sheet (MLGS). And then Fe_2O_3 nanoparticles are densely deposited on the surface of 3D-carbon. This composite shows a more stable cycling and rate performance.

2. Experimental

2.1. Synthesis of 3D-carbon material

The schematic synthesis process of Fe_2O_3/3D-carbon composite is shown in Figure 1. 20 mg expanded graphite was added into mixed solution of 8 ml DMF and 2 ml distilled water and ultrasonicated 2 h to obtain MLGS. Then, 100 mg Co $(CH_3COO)_{2.4H2O}$ was added into the above suspension and stirred 5 min to obtain uniform solution. Subsequently, the suspension was put into a 30 ml Teflon-lined autoclave and sealed. Then the stainless steel autoclave was heated at 100 ℃ for 2 h. The autoclave was subsequently cooled to ambient temperature. The precipitate was collected and washed repeatedly with ethanol and distilled water by centrifugation (9000 r min^{-1}, 5min), and then dried in an oven at 60 ℃. MLGS with homogeneous Co_3O_4 nanoparticles on its surface were obtained.

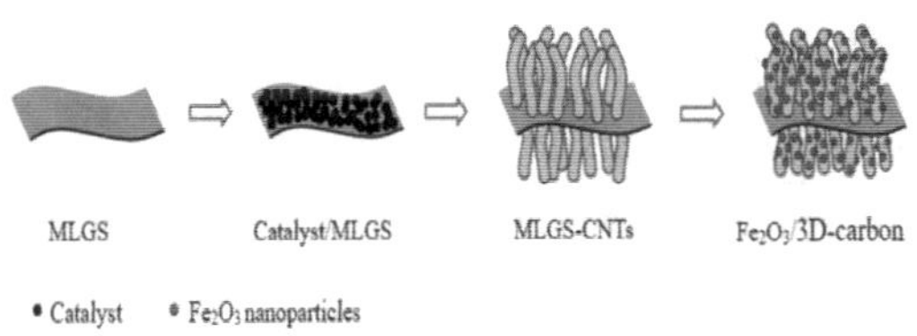

Figure 1. Schematic synthesis process of Fe_2O_3/3D-carbon composite

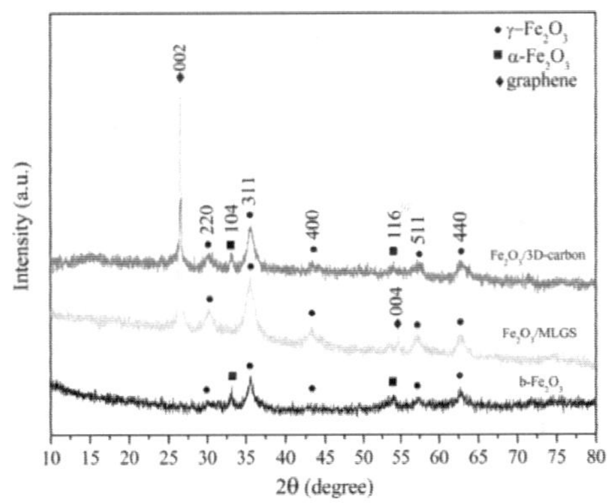

Figure 2. XRD diffraction patterns of samples

A CVD reactor was heated to 630 °C. After oxygen was cleaned up with N_2 (99.999%) at a flow rate of 200 sccm, the sample was loaded into the reactor. Then acetylene (99.95%) was introduced in with a flow rate of 40 sccm and kept flowing for 10 min. The reactor was cooled to room temperature in an N_2 atmosphere. Then the black powders were put in hydrochloric acid (36.5%) and stirred at 80 °C for 24 h to remove the catalysts. After cooling down to room temperature, the powder was washed with distilled water until the pH value of the filtrate was 7. Finally, the 3D-carbon was obtained after dried at 60 °C for 12 h.

2.2. Synthesis of Fe2O3/3D-carbon composites

IOP Conf. Series: Materials Science and Engineering **761** (2020) 012002 doi:10.1088/1757-899X/761/1/012002

20 mg 3D-carbon, 96 mg $FeCl_2 \cdot 4H_2O$ and 200 mg CH_3COONa was added into mixed solution of 8 ml DMF and 2 ml distilled water, then the mixed solution was stirred 5 min to obtain uniform solution. Subsequently, the suspension was put into a Teflon-lined autoclave and sealed at room temperature. Then the stainless steel autoclave was heated at 100 °C for 2 h. After the autoclave was cooled to ambient temperature, the precipitate was collected by centrifugation (9000 r min^{-1}, 5min) and washed repeatedly with water and ethanol. Finally, Fe_2O_3/3D-carbon composite was collected after dried at 60 °C for 24 h.

For comparison, bare Fe_2O_3 (b-Fe_2O_3) was synthesized in the same conditions but without addition of 3D-carbon. Fe_2O_3/MLGS was synthesized by replacing 3D-carbon with MLGS.

2.3. Material Characterization

Samples were investigated with X-ray diffraction (XRD, Shimadzu XRD-6000, Cu Kα radiation, 1.5406 Å). The scan range was from 10 °C to 80 °C. Morphologies of samples were examined by scanning electron microscopy (SEM, Hitachi S-4800) at 5 kV and transmission electron microscopy (TEM, Philips 2130) at 120 kV.

2.4. Electrochemical measurements

Electrochemical tests were carried out using CR2032 coin-type cell with pure lithium foil as the counter electrode. The working electrodes were fabricated by mixing as-prepared sample: active carbon: poly (vinyldifluoride) (PVDF) with a weight ratio of 70:15:15 in N-methylpyrrolidone (NMP) solvent and then pasting onto copper foils. The lithium foil was used as counter/reference electrodes and the electrolyte was a solution of 1M $LiPF_6$ in ethylene carbonate (EC)/dimethyl carbonate (DMC) (1/1, w/w). The cells were tested on a NEWARE multi-channel battery test system with galvanostatic charge/discharge (GCD) technique in a voltage range of 0.01-3.0V. CHI660e electrochemical working station was employed for cyclic voltammetry (CV) measurement. Electrochemical impedance spectroscopy (EIS) measurement was also carried out on CHI600E chemistry workstation by applying a perturbation voltage of 10 mV in a frequency range from 0.01 to 100 KHz.

3. Results and discussion

3.1. Characterization of materials

To confirm the crystal structures of the obtained composites, the XRD patterns of three samples were performed. The XRD patterns of sample Fe_2O_3/3D-carbon, Fe_2O_3/MLGS and b-Fe_2O_3 are showed in Figure 2. The peaks at 26.602° and 54.715° correspond to the graphitic lattice plane (002) and (004) of the MLGS and CNT. The peaks at 30.3°, 35.7°, 43.3°, 57.3°, 62.9° match well with the characteristic diffraction plane (220), (311), (400), (511), (440) of γ-Fe_2O_3 phase (JCPDS 39-1346). The weak peaks at 33.1, 54.1 belong to the (104), (116) plane of α-Fe_2O_3 (JCPDS 24-0072). Little α-Fe_2O_3 phase is present may due to the faster oxidation of Fe^{2+} [4].

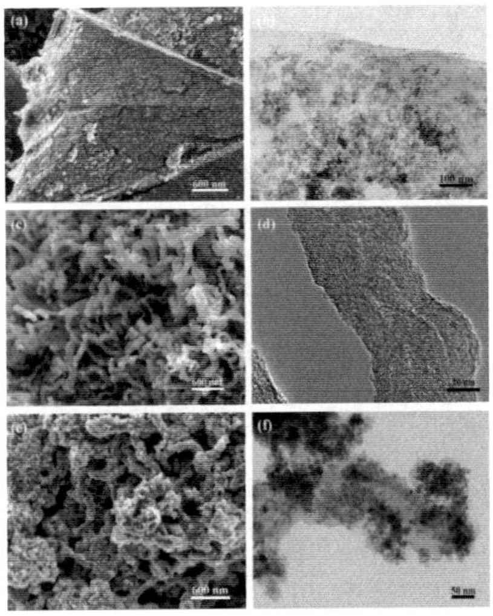

Figure 3. SEM and TEM images of samples, (a, b) Fe_2O_3/MLGS; (c, d)3D-carbon; (e, f) Fe_2O_3/3D-carbon

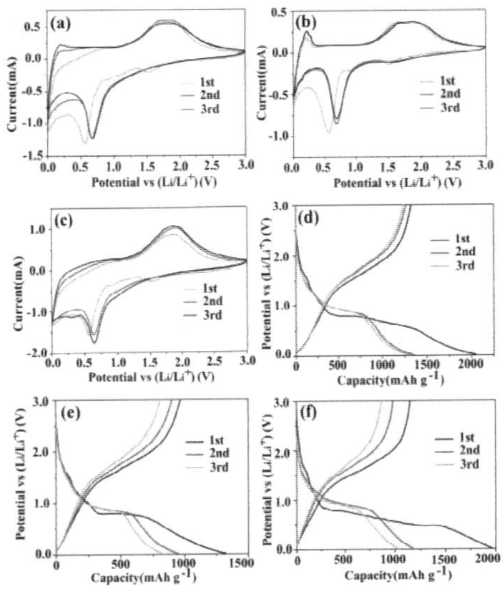

Figure 4. CV curves of (a) Fe_2O_3/3D-carbon, (b) Fe_2O_3/MLGS, (c) b-Fe_2O_3; GCD curves of (d) Fe_2O_3/3D-carbon, (e) Fe_2O_3/MLGS, (f) b-Fe_2O_3.

The morphologies and microstructures of the products were investigated by SEM and TEM, as shown in Figure 3. It is obvious that the Fe_2O_3 nanoparticles with the size of about 5 nm uniformly cover the surface of MLGS (Figure 3a and b). Figure 3c and 3d show that CNTs with a diameter of around 50-70 nm are grown on the MGLS and form a continuous electron conducting network structure. Figure 3e and 3f shows that the Fe_2O_3 nanoparticles are homogeneously dispersed on the CNTs with a size at the rage of 5-10nm after this synthesis process.

3.2. Electrochemical performance

Figure 4 displays the initial three cyclic voltammetry (CV) curves of three electrodes at a scan rate of 0.1 mV s^{-1} in the voltage window of 0.01-3.0 V. The peak at 0.56 V in the first cathodic branch could be ascribed to the reduction of Fe (III) to Fe (0), as well as the formation of solid electrolyte interphase (SEI) layer. In the following discharge cycles, the reduction peak at 0.56 V shifts to 0.70 V. In subsequent cycles, the cathodic peaks tend to stable, confirming good reversibility and high columbic efficiency of the samples. In the charging process, the two anodic peaks are observed at 1.65 V and 1.85 V, which associated with the reversible multistep oxidation of Fe (0) to Fe (II) and further oxidation to Fe (III). The peak below 0.5 V is due to lithium intercalation and de-intercalation in the graphitic layers [5], which are not observed in b-Fe_2O_3 sample.

The typical GCD curves for Fe_2O_3/3D-carbon, Fe_2O_3/MLGS and b-Fe_2O_3 samples at 50 mA g^{-1} are presented in Figure 4d, 4e and 4f. It can be observed from Figure 4f that the first discharge platform of the Fe_2O_3/3D-carbon is at 0.56 V and the subsequent discharge platform is changed to near 0.7 V, corresponding to the position of the reduction peak shown in the CV curve Figure 4c. The first discharge and charge capacity is 2050 mAh g^{-1} and 1327 mAh g^{-1} for Fe_2O_3/3D-carbon, which is higher than theoretical capacity. The great capacity loss in the initial cycles of electrode is due to the formation of SEI film, the irreversible storage of lithium storage in carbon defects, and incomplete conversion reactions in the Fe_2O_3[6]. The second and third charge/discharge curves of Fe_2O_3/3D-carbon are close to each other, which indicate the relative good stability of the Fe_2O_3/3D-carbon. The reversible specific capacities of Fe_2O_3/MLGS and b-Fe_2O_3 samples have dropped a lot between the third and second cycles, showing the relative poor stability.

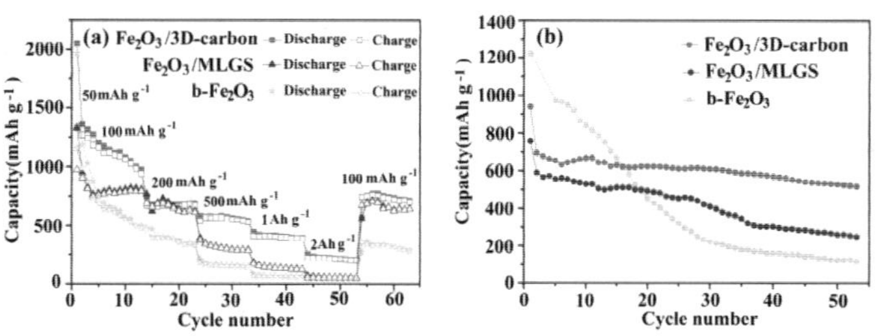

Figure 5. (a) Rate performance, (b) Cycle performance at a current density of 500 mA g^{-1} of Fe_2O_3/3D-carbon, Fe_2O_3/MLGS and b-Fe_2O_3 at different current densities.

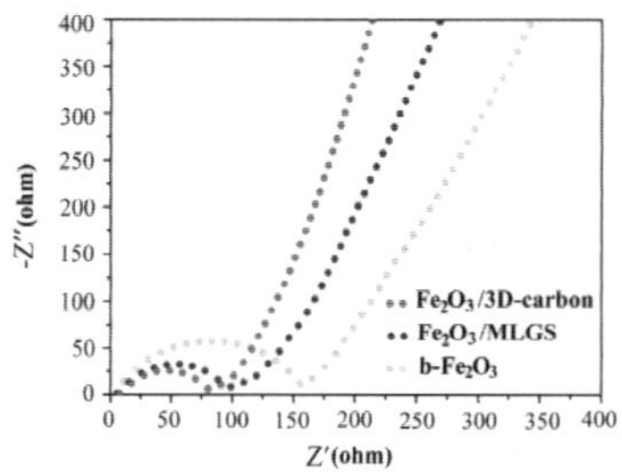

Figure 6. Nyquist plots of Fe_2O_3/3D-carbon, Fe_2O_3/MLGS and b-Fe_2O_3 electrode

Figure 5a shows the rate dependence on the capacity performance of Fe_2O_3/3D-carbon, Fe_2O_3/MLGS and b-Fe_2O_3 electrodes. Under the current densities of 0.1, 0.2, 0.5, 1 and 2 A g^{-1}, the reversible capacities are 1025, 678, 547, 394 and 205 mAh g^{-1} for Fe_2O_3/3D-carbon, 817, 630, 291, 128 and 60 mAh g^{-1} for Fe_2O_3/MLGS, and 525, 342, 154, 70 and 23 mAh g^{-1} for b-Fe_2O_3. Figure 5b shows the cycling performance of three electrodes at high current density of 500 mA g^{-1}. Specific capacity 528 mAh g^{-1} is maintained after 50 cycles for Fe_2O_3/3D-carbon, while 226 mAh g^{-1} is for Fe_2O_3/MLGS and only 98 mAh g^{-1} is for b-Fe_2O_3. The results show the 3D-carbon supported Fe_2O_3 has higher performance than that of MLGS supported and bare Fe_2O_3 nanoparticles.

EIS measurements were carried out to study the transport kinetics of electrons and ion ability in the electrodes. The Nyquist plots of Fe_2O_3/3D-carbon, Fe_2O_3/MLGS and b-Fe_2O_3 electrodes are shown in Figure 6. A typical semicircle in the high-medium frequency region can be ascribed to the charge-transfer resistance. The radius for the Fe_2O_3/3D-carbon is smaller than that for both Fe_2O_3/MLGS and b-Fe_2O_3, implying that Fe_2O_3/3D-carbon composite electrode exhibits faster charge-transfer and smaller electrochemical reaction resistance than that of Fe_2O_3/MLGS and Fe_2O_3. At low frequency region, the steeper tail of the Fe_2O_3/3D-carbon electrode indicates lower ion diffusion resistance compared to the Fe_2O_3/MLGS and Fe_2O_3 electrode.

4. Conclusions

In summary, Hydrothermal method was developed to deposit homogeneous Fe_2O_3 nanoparticles in size of about 5-10 nm on 3D-carbon based on π-π interaction. Compared with the Fe_2O_3/MLGS and bare Fe_2O_3, the Fe_2O_3/3D-carbon delivers a higher reversible capacity, and exhibits better cycling performance. Meanwhile, Fe_2O_3/3D-carbon exhibits faster charge-transfer and smaller electrochemical reaction resistance. The excellent electrochemical performance of Fe_2O_3/3D-carbon can be attributed to their 3D conductive network and porous structure.

5. References

[1] Meng J K, Fu L, Liu Y S, et al. Gas-liquid interfacial assembly and electrochemical properties of 3D highly dispersed α-Fe_2O_3@graphene aerogel composites with a hierarchical structure for applications in anodes of lithium ion batteries [J]. Electrochimica Acta, 2017, 224:40-48.

[2] Cheng F, Huang K, Liu S, et al. Surfactant carbonization to synthesize pseudocubic α-Fe_2O_3/C nanocomposite and its electrochemical performance in lithium-ion batteries [J]. Electrochimica Acta, 2011, 56(16): 5593-5598.

[3] Wu M, Chen J, Wang C, et al. Facile Synthesis of Fe_2O_3 Nanobelts/CNTs Composites as High-performance Anode for Lithium-ion Battery[J]. Electrochimica Acta, 2014, 132(3): 533-537.

[4] Qiu X, XU J, et al. Preparation of Uniform Magnetic Fe_2O_3 Nanoparticles@multi-layer Graphene Composites with Complexation Method [J]. Journal of Inorganic Materials, 2016, 31(12):1335-1340.

[5] Wu Y, Gong M, Lin M C, et al. 3D Graphitic Foams Derived from Chloroaluminate Anion Intercalation for Ultrafast Aluminum-Ion Battery[J]. Advanced Materials, 2016, 28(41): 9218-9222.

[6] Zeng, H. Zhao, P. Lv, Z. Zhang, J. Wang, Q. Xia, Electrochemical properties of iron oxides/carbon nanotubes as anode material for lithium ion batteries, J. Power Sources 274 (2015) 1091–1099

IOP Conf. Series: Materials Science and Engineering **761** (2020) 012003 doi:10.1088/1757-899X/761/1/012003

Preparation And Electrochemical Performance of Nano Si@Short Hard Carbon Fiber Composites as Anodes for Lithium Batteries

J WANG[1*], X Y LIU [1,2*] and A XING[1]

[1]School of Metallurgy and Material Engineering,Chongqing University of Science and Technology,Chongqing,China
[2]Shenzhen Key Laboratory of Nanobiomechanics, Shenzhen Institutes of Advanced Technology, China Academy of Sciences, Shenzhen, Guangdong, China

Abstract. The bamboo paper was soaked by the silica sol.After drying, the nano-silicon@short hard carbon fiber was prepared by vacuum high-temperature carbonization treatment followed by grind, which was one-dimensional composite. The structure and morphology of the composite material were characterized by X-ray diffraction and Scanning electron microscopy (SEM). The results show that the diameter of the short hard carbon fiber is about 2~8 μm and the length is about several tens of microns. The nano-silicon particles are sparsely distributed on the outer wall of the short bamboo charcoal fiber. The adhesion between nano silicon and short hard carbon fiber is strong due to the effect of thermal diffusion and Van der Waals' force. The prepared nano-silicon@short hard carbon fiber composite was used as a negative electrode material for lithium battery for charge and discharge test. When the current density was 100 mA.g^{-1}, the capacity was only attenuated by 6.54% after 50 cycles. Meanwhile, the composite also exhibits well rate performance.
Keywords. one dimensional composites; short hard charcoal fiber; anode material; lithium ion batteries

1. Introduction

The lithium-ion battery has many advantages, such as small size, high energy density, high power, long term using and good safety, which is ideal for energy storage [1-3]. Cathode materials play an important role in the performance of lithium-ion batteries [4-6]. The cathode material should have as low as possible electrode potential, high Li+ mobility, high reversibility of Li+ embedding and de-embedding, good conductivity and thermodynamic stability. At present, graphite is the commercialized cathode material for lithium-ion batteries [7-9]. Compared with graphite, hard carbon cathode has higher theoretical discharge capacity [10], faster charge and discharge speed, lower carbonization temperature and less energy loss, so it has become one of the research hotspots.

Bamboo is planted in large quantities in China, with abundant sources and low prices. In this paper, bamboo pulp paper was used as raw material; hard carbon fiber was prepared by carbonizing at 800 ℃ [11], then ground into short bamboo carbon fiber. It was used as cathode material for lithium-ion batteries to test its charge-discharge performance. The results show that it has lithium storage capacity and good cycling performance. In addition, one-dimensional composites of nano-silicon @ short bamboo carbon fiber were prepared by a simple method. It was found that the electrochemical properties were greatly improved compared with pure short-cut bamboo-carbon fibers.

2. Experiment section

2.1. Raw Materials and Instruments
The purity of nano-silica powder (particle size 30 nm) used in this experiment is 99.9%. It was purchased from Zhongxin New Material Company. Bamboo pulp paper was marketed as Fanya brand bamboo pulp paper. Sodium carboxymethyl cellulose (CMC-Na) and styrene-butadiene rubber (SBR) were purchased from Cyber electrochemical material network. Analytical pure anhydrous ethanol was purchased from Chengdu Cologne Chemical Co., Ltd.

2.2. Material preparation
Bamboo paper was soaked in deionized water for 12 hours, taked out, and dried in an oven at 80℃ for 8 hours. Placing the treated bamboo paper into a vacuum carbon tube furnace and carbonizing at 800 ℃ in a vacuum environment for 8 hours to get bamboo carbon fibers. The bamboo fiber was grinded for 1 hour with agate mortar to obtain short hard carbon fiber.
Nano silicon/ethanol sol with the mass percentage concentration of 0.8% was prepared by ultrasonic dispersion. Bamboo paper was soaked in deionized water for 12 hours, taked out, and dried in an oven at 80℃ for 8 hours. The bamboo paper was soaked in nano silicon/ethanol sol in an oven at 50 ℃ for 30 minutes, taked out, and dried in a vacuum oven at 80 ℃ for 3 hours. The bamboo paper with nano-silicon was carbonized in a vacuum carbon tube furnace at 800 ℃ for 8 hours to obtain the nano-silicon @ bamboo carbon fibers composite material. The heating rate is 8℃/min. Grinding the nano silicon @ bamboo carbon fibers for 1 hour with an agate mortar to obtain the nano silicon @ short hard carbon fiber composite material.

2.3. Battery assembly and electrochemical measurements
Active substances, sodium carboxymethylcellulose (cmc) and styrene butadiene rubber (SBR) are ground and prepared by using deionized water as a dispersant according to a mass ratio of 90: 5: 5, and are uniformly coated on copper foil by a scraper coating method after being ground uniformly. The copper foil is baked in an oven at 80 ℃ for 12 hours and then is punched by a punching machine to prepare a electrode with a diameter of 14 mm. A button cell (2016) was assembled in an argon filled glove box (Mikrouna (China) Co., Ltd., Super (1220/750/900),) filled with argon, in which the concentration of oxygen and water vapor was lower than 0.01 PPm. The electrolyte was 1 mol/L LiPF6 solution (the solvent was a mixed solvent of vinyl carbonate, methyl ethyl carbonate and diethyl carbonate with a volume ratio of 1: 1: 1). Celgard 232 was used as the separator; Li foil was used as both the counter and reference electrodes.
The cells tested by LANHE CT2001A battery-test system (Wuhan LAND electronics Co., Ltd.) galvanostatic charge-discharge tests at various current densities in the voltage range of 0.01-3.0 V (versus Li/Li+).

3. Results and discussions

3.1. Morphology and structure
Figure 1 is an SEM photograph of the produced chopped hard carbon fiber. It can be seen from Figure 1(a) that the material is chopped fiber with a diameter of about 2-8 μm and a length of about tens of μm. The length of the fiber can be controlled by controlling the grinding time and strength. It can be seen from Figure 1(b) that there are irregular folds and cracks on the surface of the sample, which may be caused by cracking reaction at high temperature. These cracks and folds increase the specific surface area of the material.
Figure 2 is an xrd detection result diagram of chopped hard carbon fibers. From the diagram, it can be seen that the chopped bamboo charcoal fibers are amorphous structures, and the characteristic peak around 23° corresponds to the (002) crystal plane of carbon, which is formed by parallel stacking of

IOP Conf. Series: Materials Science and Engineering **761** (2020) 012003 doi:10.1088/1757-899X/761/1/012003

some graphitized sheets in the carbon material; The characteristic peak around 43° corresponds to the (100) crystal plane of the carbon material, which corresponds to the hexagonal lattice structure formed by the interaction of sp3 hybridization atoms in the carbon material [12].

Figure 3 is a SEM image of nano-silicon short hard carbon fiber with different magnification. The fluffy material attached to the short bamboo carbon fiber is nano-silicon. The distribution of nano-silicon on the bamboo charcoal fiber is sparse, which helps alleviating the volume expansion effect of nano-silicon during charging and discharging process. In addition, nano-silicon has a slight agglomeration phenomenon. Because the composite material has been under heat treatment at 800℃, thermal diffusion can strengthen the bonding strength between nano-silicon and short carbon fibers. Besides, the surface of nano-silicon will inevitably have some oxidation, and the oxygen atoms can be connected with oxygen-containing groups on the surface of short-cut bamboo charcoal fibers through vander Waals force, which also enhances the bonding strength between nano-silicon and short bamboo charcoal fibers. This is beneficial for the improvement of material cycling performance.

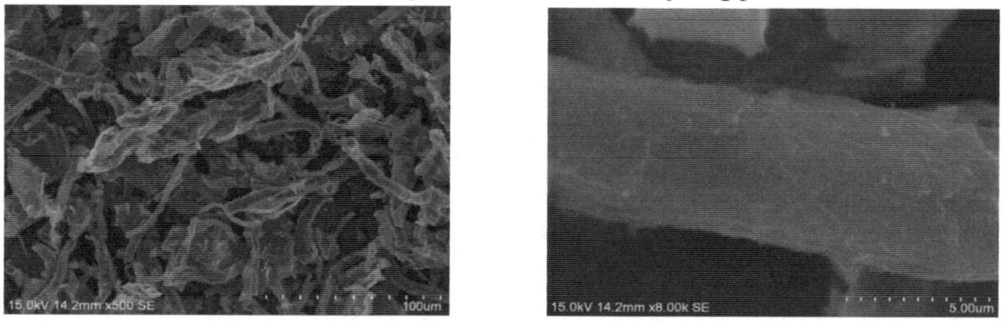

(a) (b)

Figure 1. SEM images of short hard carbon fiber

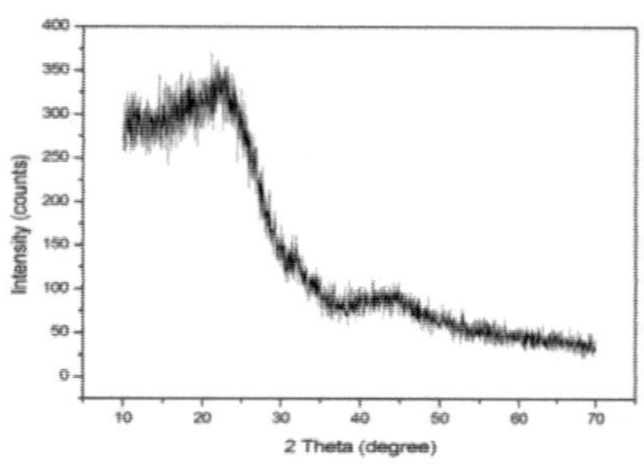

Figure 2. XRD pattern of short hard carbon fiber

Figure 3. SEM images of nano Si@short hard carbon fiber Composites

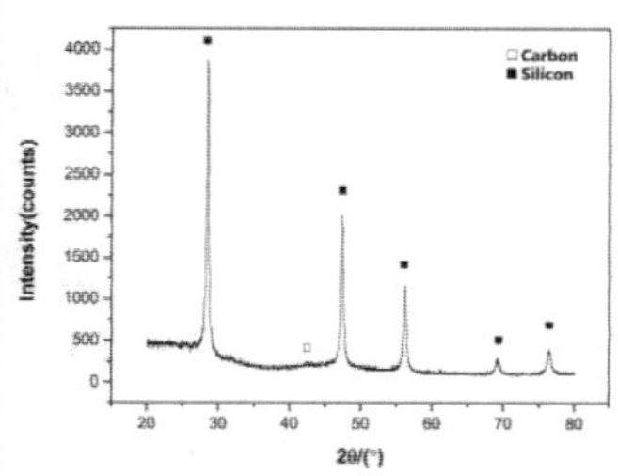

Figure 4. XRD pattern of nano Si@short hard carbon fiber Composites

Figure 4 is the XRD test chart for Nano-silicon@ short bamboo carbon fiber composites. The diffraction peak in crystal silicon is very clear, but it is very weak in bamboo carbon. This is due to that the bamboo carbon is amorphous carbon and amorphous, and it has a very week diffraction peak intensity, with the maximum intensity of about 350. Nano-silicon is a crystal with high intensity of diffraction peak, with a maximum intensity of about 400. Therefore, the diffraction peak of bamboo carbon is relatively weak when these two groups of peaks are in the same chart. In the graph, it can be observed that, in the range of 20 - 26°, the map base is relatively higher, which is caused by amorphous bamboo carbon.

3.2. Electrochemical performance analysis

IOP Conf. Series: Materials Science and Engineering **761** (2020) 012003 doi:10.1088/1757-899X/761/1/012003

Figure 5 shows the initial charge-discharge curve of the sample with a current density of 30 mA.g-1 and a charge-discharge voltage range of 0.01~1 V. It can be seen from the diagram that the initial discharge specific capacity of the short bamboo carbon fiber is 607 mAh.g-1. The initial charge specific capacity is 239 mAh.g-1, and the first charge-discharge efficiency is 39%. It has been reported in the literature that the H/C of biomass hard carbon material carbonized at low temperature is higher, which leads to a lower first charge-discharge coulombic efficiency. The initial discharge specific capacity of nano-silicon@ short bamboo carbon fiber composites is 1267 mAh.g-1, the first charge specific capacity is 764mAh.g-1, and the first charge-discharge efficiency is 60%. The introduction of nano-silicon improves the first charge-discharge coulomb efficiency of anode materials. Compared with pure short bamboo carbon fiber, the introduction of high-capacity nano-silicon [13-15] is the fundamental reason for the improvement of electrochemical capacity of nano-silicon@ short bamboo carbon fiber composites. In the first discharge curve, there is an obvious voltage platform around 0.75 V, which corresponds to the formation of SEI film.

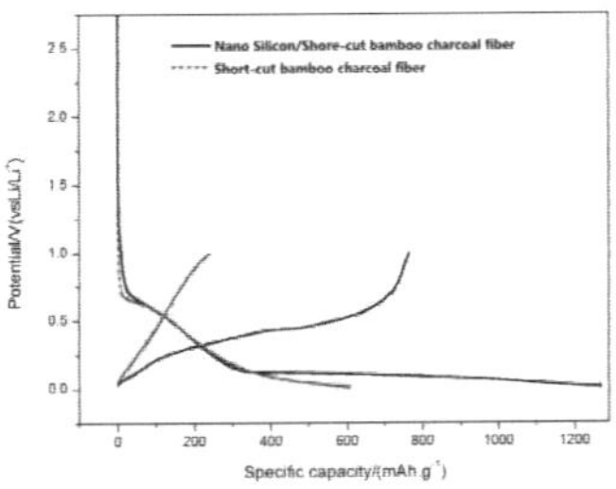

Figure 5. Charge/discharge profiles of sample at the current density of 30 mA.g-1 for the first cycles

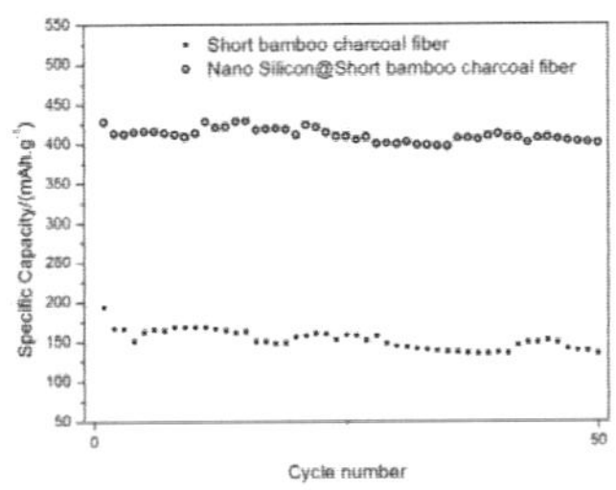

Figure 6. Cycling performance of the sample at the current density of 100 mA·g $^{-1}$

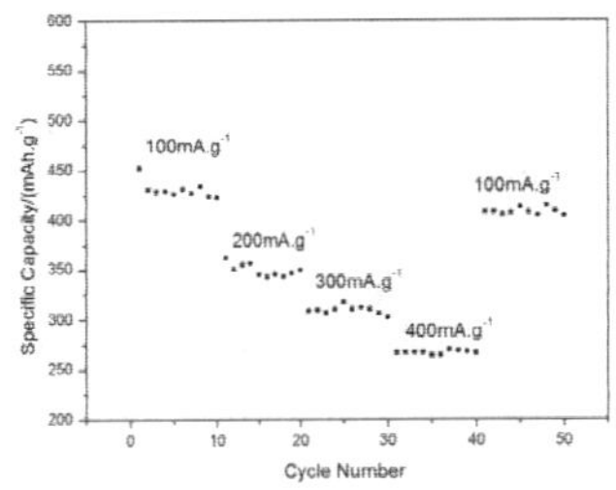

Figure 7. Rate performance of composites

Figure 6 is a graph showing the charge-discharge cycle performance of the sample at a current density of 100 ma g $^{-1}$. It can be seen from the graph that the cycle performance of chopped bamboo charcoal fiber is stable, but its capacity is not ideal and the first discharge specific capacity is only 193 mah g $^{-1}$. After nano-silicon is compounded on the chopped bamboo charcoal fiber, the capacity is obviously improved, the first discharge specific capacity reaches 428 mah g $^{-1}$, which is 121.76% higher than that of the chopped bamboo charcoal fiber and the cycle stability is not obviously reduced; after 50 cycles, the discharge specific capacity is still maintained at 400 mah g $^{-1}$, which is only reduced by 6.54%. Pure silicon active materials are accompanied by drastic volume changes [16] in the cycle process, which will lead to the destruction of the electrode structure, thus its capacity decays rapidly and its cycle performance is poor. In the nano silicon @ chopped bamboo charcoal fiber composite material, nano silicon is sparsely embedded in the chopped bamboo charcoal fiber, leaving enough space to relieve the volume change of nano silicon during charging and discharging, so the cycle performance of the composite material is not obviously reduced; In addition, the better connection between nano-silicon and chopped bamboo charcoal fibers can also alleviate the damage of electrode structure caused by the volume expansion of nano-silicon.

Figure 7 is a magnification performance chart of nano silicon/chopped hard carbon fiber composite material. It can be seen that when the current density gradually increases from 100mA.g $^{-1}$ to 400mA.g $^{-1}$, the specific discharge volume of the composite material decreases, but when it returns to 100mA.g $^{-1}$ again, the specific discharge capacity of the material basically returns to the original level. When hard carbon and silicon are respectively used as active materials of lithium battery cathode materials, their multiplying power performance is poor. However, when the two are combined together, nano silicon has higher lithium storage capacity and hard carbon fiber has excellent conductivity. The synergistic effect of the two improves the multiplying power performance of the composite material.

IOP Conf. Series: Materials Science and Engineering **761** (2020) 012003 doi:10.1088/1757-899X/761/1/012003

4. Conclusion

Short hard carbon fiber is successfully made from bamboo paper in the way of high temperature heat treatment in vacuum in this research. The preparation method for biomass-based short bamboo carbon fiber is simple and the raw material supply is wide. Bamboo paper is used as the template and soaked in silica sol. In the way of high temperature heat treatment in vacuum, compound material of nano-silicon @ short hard carbon fiber is prepared. The introduction of high-capacity nano-silicon improves the lithium storage capacity of active materials. Nano-silicon attaches on short-cut bamboo charcoal fiber sparsely. When it is made as pole pieces, nano-silicon is covered and buried into carbon fiber. This structure relieves the volume expansion in nano-silicon cycling and it is good for the improvement of cycling performance of compound materials. The preparation method in this research is simple and cost-effective. In such an era when the problems of environment protection and energy-saving are getting more and more serious, this method is of great research significance.

5. References

[1] Wen L, Li F and Cheng H-M 2016 Advanced Materials **28** 4306.
[2] PENG Y T and LO C T 2015 Journal of the Electrochemical Society **162** A1085.
[3] Mahmut D, Ozkan Y, Lu Y, Fang X M, Han J, Huseyin K and Zhang X W 2016 Electrochimica **169** 52.
[4] ZHOU B Y, ZHANG C Y,WU C H and LIU H 2016 CHINESE J INORG CHEM **32** 811
[5] WANG M M, PAN J and CHENG Y J 2016 Materials Review **30** 16.
[6] Tao W, Xu M L, Zhu Y R, Zhang Q Y and Yi T F 2017 SCIENCE CHINA Materials **60** 728.
[7] TIAN A H, WEI W, QU P, XIA Q P and SHEN Q 2017 Acta Phys. -Chim. Sin. **33** 1621.
[8] TANG F L, LEI J L, CUI Z Y, Jian O, LIU G and ZHAO L Z 2015 Trans. Nonferrous Met. Soc. China. **25** 4046.
[9] GENG K M, WU J J, GENG H B, HU Y Y, QU G L, PAN Y, ZHENG J W and GU H W 2016 CHINESE JOURNAL OF INORGANIC CHEMISTRY **32** 1495.
[10] Xie Q S, Wang L and Liu P 2016 CARBON TECHNIQUES **35** 43.
[11] LI Y L, LIU R F, ZHOU Y, GUO H Y, HE L and QIU J S 2017 Materials Review **31** 236.
[12] Li C Q, Yang D J,Xi Y B, Qin Y L and Qiu X Q 2018 Chemical Research in Chinese Universities **39** 2725.
[13] YANG K, HOU C and SONG X Y 2018 Journal of Inorganic Materials **33** 1065.
[14] SHEN X H, FAN R J, TIAN Z Y, ZHANG D P, CAO G L and SHAO L 2017 JOURNAL OF THE CHINESE CERAMIC SOCIETY **45** 1530.
[15] TAO H C, WANG M S and FAN L Z 2013 JOURNAL OF THE CHINESE CERAMIC SOCIETY **41** 1046.
[16] Wu B Z, Wu F Z, Jin H X, Lu J T and Chen J B 2018 RARE METAL MATERIALS AND ENGINEERING **47** 2600.

Acknowledgements

This work was financially supported by the National Natural Science Foundation of China (51472037), the Foundation from Shenzhen Science and Technology Innovation Committee (JCYJ20170818160815002), Fostering Foundation for Key Projects from Chongqing University of Science and Technology (CK2015Z20).

IOP Conf. Series: Materials Science and Engineering **761** (2020) 012004 doi:10.1088/1757-899X/761/1/012004

Study on the Preparation of Lifepo4 by Hydrothermal Method

Y Liu, G Y Luo, Y J Gu[1], F Z Wu, Y Mai and X Y Dai
Guizhou Province key Laboratory of Metallurgical Engineering and Process Energy Saving, College of Material and Metallurgy, Guizhou University, Guizhou, Guiyang, 550025, China
[1] Tel.: +86-851-83627683; fax: +86-851-83627683; yijing-gu@163.com

Abstract. A simple hydrothermal process combined with carbon coating is attempted to prepare nano particle $LiFePO_4$ cathode materials for Li-ion batteries. A carbon coating process with glucose is used to make $LiFePO_4$/C composites. The structure, morphology and electrochemical properties of the materials are characterized by X-ray diffraction (XRD), scanning electron microscopy (SEM) and constant current charge/discharge. The X-ray diffraction pattern indicates that the pure phase $LiFePO_4$ material could be prepared at different reaction temperatures, and the pure phase $LiFePO_4$ material could not be prepared with the reaction time of 6 h. SEM test shows that the dispersion of $LiFePO_4$ particles become better with the increase of reaction temperatures and the morphology is more regular with the increase of reaction temperature. But when the reaction temperature increases to 200 °C, $LiFePO_4$ particles reaggregate severely. As the reaction time increases, the particles of the sample gradually grow. The electrochemical performance indicates that the sample synthesized at 180 °C for 10 h has the highest specific capacity of 139.5 mAh·g^{-1} at 0.1 C. This study shows that proposed process can be a potential promising way to prepare high performance $LiFePO_4$ cathode materials.

1. Introduction

The olivine lithium iron phosphate($LiFePO_4$) has been regarded the most promising lithium ion batteries cathode material because of the advantages of excellent energy density, good cycling performance, high operating voltage, low-cost and environment friendly [1-4]. Nowadays $LiFePO_4$ has been applied in all areas of our life [5]. However, the low electronic and ionic conductivity of single phase $LiFePO_4$ result in poor electrical properties and awful thermostability, especially the slow lithium ion diffusivity of pristine $LiFePO_4$ at lower temperature which severely restrict its extensive application in practice [6-8]. In order to improve the properties of $LiFePO_4$, a lot of researches have been done to overcome the defects of lithium iron phosphate by decreasing particle size, optimizing the microcosmic morphology, doping other metal ions [9-11]. The morphology of $LiFePO_4$ is the key factor to achieve high specific capacity and cycling performance. Choosing appropriate preparation method can not only improve microstructure, but also make the powder particle size reach nano-sized [12, 13]. Moreover, the sample particle has the advantages of good uniformity and dispersity which shortens lithium ion diffusion as well as improves electron transfer efficiency and then improves the performance of $LiFePO_4$ cathode material [14].

High temperature solid state reaction method is the conventional approach to synthesis $LiFePO_4$ due to the advantages of simple synthesis process and convenient control [15-17]. However, the product particle size is thick and large, and the most serve defects are the lower purity and poor uniformity [18]. Compared with state reaction method, hydrothermal synthesis method has the advantages of high crystallinity, uniform particle, simple process, low energy consumption and short period [19, 20].

In this paper, we select hydrothermal synthesis method to prepare $LiFePO_4$ cathode material, and investigate the effect of temperature and time on the performance of $LiFePO_4$ cathode material.

2. Experimental

2.1. Sample preparation

The sample powders were synthesized via hydrothermal synthesis method, using $LiOH{\cdot}H_2O$, $FeSO_4{\cdot}7H_2O$, and H_3PO_4 as raw materials. They were dissolved into deionized water according to the stoichiometric amounts of $LiFePO_4$ and stirred at room temperature for several minutes to form $LiFePO_4$ precursor. In particular, the above process was conducted in Ar atmosphere. The obtained precursor solution was transferred into a autoclave and kept at 140 °C, 160 °C, 180 °C and 200 °C for 10 h. Then the solution was washed with deionized water for several times and filtered. Subsequently, the filtered powders were dried in the oven at 80°C and then milled in an agate mortar. Then was heated at 650 °C for 6 h which filled with Ar gas. The $LiFePO_4$ powders were synthesized. Choose appropriate temperature and set the hydrothermal reaction time to 6 h, 8 h, 10 h, 12 h, 15 h and repeated above process.

2.2. Phase structure and morphology characterization

The crystal structures of all $LiFePO_4$ were analyzed by XRD (D/Max2500, Japan). The microstructure of as-prepared samples was examined by SEM (FESEM, JEOL, 6300/7001).

2.3. Electrochemical characterization

$LiFePO_4$ was used as cathode electrode active material, acetylene black as a conductive agent, polyvinylidene fluoride (PVDF) as a binder. According to the mass ratio of $LiFePO_4$: acetylene black : PVDF is 8: 1: 1. NMP is added as solvent to make it fully mixed. Then the slurry was coated on the foil and dried in vacuum at 100 °C for 24 h. Sliced and assembled the button cell in a vacuum glove box. Lithium metal was used as the negative electrode and Celgard 2400 (polypropylene) was used as the separator. Constant current charge and discharge experiments were tested from 2.5 V to 4.5 V by using Blue Electronic battery test system.

3. Results and discussion

In the process of synthesizing $LiFePO_4$, the temperature and time of hydrothermal reaction are important factors. When the reaction temperature is low, the iron and lithium dislocations are generated easily in the crystal lattice, which affects the electrochemical properties. When the temperature is raised, the precipitation and grain growth of the crystal during hydrothermal process can be enhanced. Figure 1 shows the XRD patterns of the $LiFePO_4$ particles obtained at various reaction temperatures at a fixed reaction time of 10 h. It can be seen from Figure 1 that the XRD patterns of all the samples are consistent with the JCPDS (81-1173) standard, and no diffraction peaks of any impurity phases are detected in each sample, indicating that high-purity $LiFePO_4$ samples are prepared successfully. The peaks of the sample obtained at 180 °C are sharper, indicating that the crystallinity of the material is better and the degree of crystallization is higher when the hydrothermal temperature is 180 °C.

As shown in Figure 2, all of the $LiFePO_4$ particles are tiny cuboids over the entire temperature range. The particle size prepared at 200 °C is uneven obviously, and some large particle size can be seen. That may be because the temperature is higher to affect the nucleation reaction of the material. The other samples are monodisperse at lower temperatures, but the sheet-like $LiFePO_4$ nanoparticles with a smaller size appear at 180 °C. With the increase of the reaction temperatures, the dispersibility of the $LiFePO_4$ particles become better and the morphology is more regular until 180 °C. The $LiFePO_4$ particles agglomerated when the temperature is 200 °C.

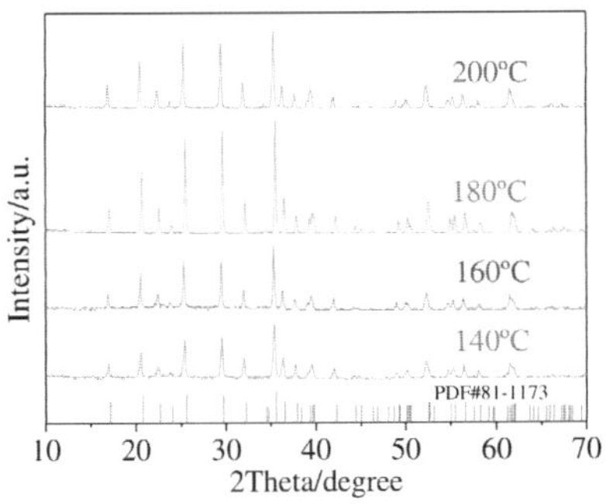

Figure 1. XRD patterns of $LiFePO_4/C$ prepared at different reaction temperatures.

Figure 2. SEM images of $LiFePO_4/C$; (a) 140 °C, (b) 160 °C, (c) 180 °C, (d) 200 °C.

Figure 3 is the first charge/discharge diagram of $LiFePO_4$ prepared at different reaction temperatures at 0.1 C rate. The charge/discharge curve platform is stable, corresponding to the two-phase transformation process of $LiFePO_4/FePO_4$. With the reaction temperatures rising from 140 °C to 200 °C, the specific capacities of samples are 28.5, 69.2, 139.5 and 111.6 $mAh·g^{-1}$.The specific capacity of the samples increase with the temperature from 140 °C to 180 °C, but the specific capacity decreases at 200 °C. The sample prepared at 180 °C has the highest specific capacity. From all of the above analysis, the comprehensive performance of the material at 180 °C is the most superior.
Figure 4 is the XRD patterns at different reaction times at 180 °C. It can be seen that the sample with reaction time of 6 h does not correspond to the standard patterns, indicating that $LiFePO_4$ cannot be synthesized. The other XRD patterns of the reaction time are consistent with the standard card. The peaks are sharp and no impurity peaks. That indicates the high-purity samples are prepared at other time. The half-peak width of the sample at 10 h is narrow, indicating that the crystallinity increases.

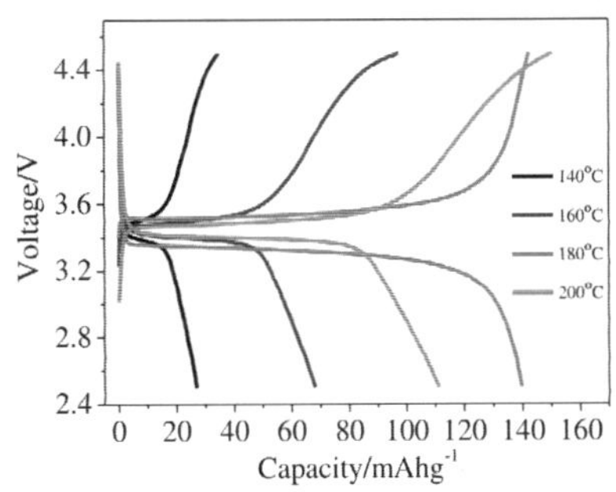

Figure 3. First charge and discharge diagram of $LiFePO_4/C$ at different reaction temperatures.

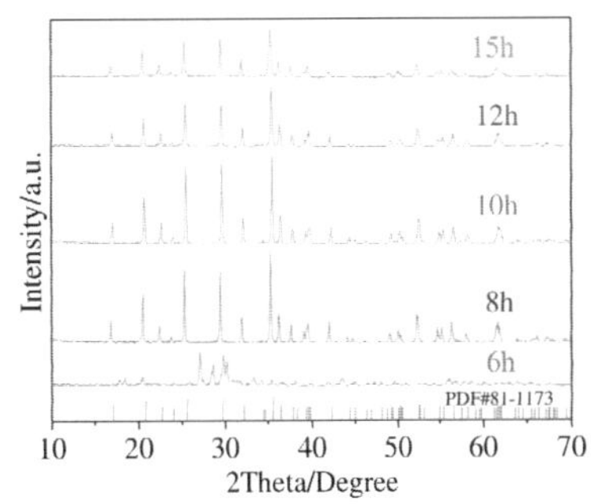

Figure 4. XRD patterns of $LiFePO_4/C$ prepared at different reaction times

The SEM images at different reaction times is shown in Figure 5. As the reaction time increases, the average particle size gradually increases. No obvious particles are observed at 6 h. This should be caused by the hydrothermal reaction time is too short to form $LiFePO_4$ which matches with the XRD. When the time is 8 h, the particle is small and agglomerates together. At 10 h, the particles have good dispersibility and become uniform with lathing shape. At 12 h, the particle size increases significantly. At 15 h, the particles start to grow and change their shapes. This is because the essence of the hydrothermal reaction is the process of dissolution-crystallization-regeneration of the material. First, the relatively large particles begin to dissolve into smaller particles, and the dispersibility is improved, and then the dissolved small particles are continuously crystallized again. As the reaction time increases, the crystallized particles undergo a regrowth process and grow into large particles.

Figure 5. SEM images of $LiFePO_4/C$ prepared at different reaction times; (a) 6 h, (b) 8 h, (c) 10 h, (d) 12 h, (e) 15 h.

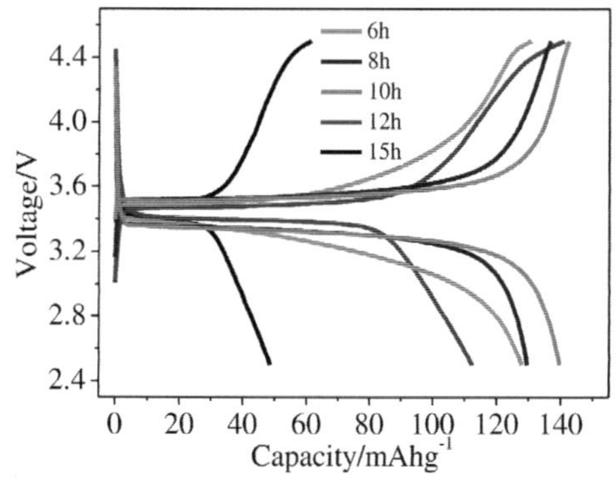

Figure 6. First charge and discharge diagram of $LiFePO_4/C$ prepared at different reaction times.

Figure 6 shows the first charge/discharge diagram of the samples at different reaction times. The specific capacities of samples at 0.1C rate are 127.9, 129.6, 139.6, 112.3 and 48.5 mAh·g^{-1}. As the reaction time increases, the specific capacity increases first and then decreases. The specific capacity of the sample at 15 h is lowest. That may be because the reaction time is too long to make the particle size large. At 10 h, the sample with a larger specific surface area leads to increase in the contact area with the electrolyte, resulting in a highest charge/discharge capacity.

4. Conclusion

$LiFePO_4$ cathode materials were prepared by simple hydrothermal method. The structure, morphology and electrochemical properties of the materials were characterized by X-ray diffraction (XRD), scanning electron microscopy (SEM) and constant current charge/discharge. The effect of reaction temperatures and times on the structure, morphology and electrochemical properties of $LiFePO_4$ were investigated. The results show that the pure phase $LiFePO_4$ nanoparticles can be prepared at a reaction temperature of 180 °C. When the reaction time is 10 h, the prepared $LiFePO_4$ has high purity and good crystallinity. Therefore, the final preparation process was determined to be 180 °C for 10 h. In particular, $LiFePO_4$ particles prepared by this hydrothermal process exhibit a reversible capacity of 139.5 mAh·g^{-1}, so the hydrothermal process is a promising strategy for industry production of $LiFePO_4$ with high performance.

5. References

[1] Zhang K, Lee J T, Li P, Kang B, Kim J H, Yi G R and Park J H 2015 *Nano. lett.* **15** 6756-63
[2] Bao L, Li L L, Xu G, Wang J W, Zhao R Y, Shen G, Han G R and Zhou S X 2016 *Electrochim. Acta* **222** 685-92
[3] Wang Y J, Zhu B, Wang Y M and Wang F 2016 *Ceram. Int.* **42** 10297-303
[4] Gong C L, Xue Z G, Ye Y S and Xie X L 2016 *J. Power Sources* **318** 93-112

[5] Sun X D and Zhang L 2018 *J. Phys. Chem. Solids* **116** 216-21
[6] Wang X L, Dong S L and H Wang H X 2018 *Ceram. Int.* **44** 464-70
[7] Wang B, Liu A, Abdulla W A, Wang D L and Zhao X S 2015 *Nanoscale* **7** 8819-28
[8] Wang X F, Feng Z J, Huang J T, Deng W, Li X B, Zhang H S and Wen Z H 2018 *Carbon* **127** 149-57
[9] Lu C H, Li W Y , Subburaj T, Ou C Y and Kumar P S 2019 *Ceram. Int.* **45** 12218-24
[10] Bao L, Xu G, Sun X L, Zeng H, Zhao R Y, Yang X, Shen G, Han G R and Zhou S X 2017 *J. Alloy. Compd.* **708** 685-93
[11] Johnson I D, Blagovidova E, Dingwall P A, Brett D L, Shearing P R and Darr J A 2016 *J. Power Sources* **326** 476-81
[12] Saroha R, Panwar A K, Sharma Y, Tyagi P K and Ghosh S 2017 *Appl. Surf. Sci.* **394** 25-36
[13] Wang Y, Feng Z H, Wang L L, Yu L, Chen J J, Liang Z and Wang R 2014 *RSC Adv.* **4** 51609-14
[14] Naik A, Zhou J, Gao C, Liu G Z and Wang L 2016 *J. Energy Inst.* **89** 21-29
[15] Oh S W, Huang Z D, Zhang B, Yu Y, He Y B and Kim J K 2012 *J. Mater. Chem.* **22** 17215-21
[16] Luo S H, Sun Y, Bao S, Li J Z, Zhang J and Yi T F 2019 *J. Electroanal. Chem.* **832** 196-203
[17] Wang X, Wang X Y, Zhang R, Wang Y and Shu H B 2018 *Waste Manage.* **78** 208-16
[18] Devaraju M K and Honma I 2012 *Adv. Energy Mater.* **2** 284-97
[19] Yin C J, Bao Z Q, Tan H, Zhou H M and Li J 2019 *Chem. Eng. J.* **372** 408-19
[20] Ziolkowska D A, Jasinski J B, Hamankiewicz B, Korona K P, Wu S H and Czerwinski A 2016 *Cryst. Growth Des.* **16** 5006-16

Acknowledgments
This work was financially supported by Guizhou Science and Technology Cooperation Project (LH [2015]7648, [2016]7440, [2017]7246), Guizhou Science and Technology Platform Talent Project [2017]5788, Guizhou University Talent Introduce Project (2014)47, the National Natural Science Foundation of China (NSFC, 51704082, 51702061), Project of the innovation group of the Guizhou Provincial Education Department (KY[2017]030), the Special funds for training outstanding young scientists and technicians in Guizhou ((2015)16).

Effect of Alloy Composition on the Optoelectronic Properties of Hydrogenated Microcrystalline Silicon-Germanium Films Deposited with Various Hydrogen Dilution

T W Li[1], Y Q Xu[1], S Y Wang[1], Y W Yu[1], Y Ma[1] and W Y Niu[1]
[1] College of Materials Science and Engineering, Shenyang Jianzhu University, Liaoning 110168, China

fighting_5@163.com.

Abstract. Hydrogenated microcrystalline silicon-germanium (μc-$Si_{1-x}Ge_x$: H) films were fabricated by plasma-enhanced chemical vapor deposition. The influence of hydrogen dilution on the optoelectronic and crystal structural properties of μc-$Si_{1-x}Ge_x$: H films were investigated upon alloy composition. Under the relatively low hydrogen dilution condition, the Si atoms of films are easier to crystallize than the Ge atoms, and the dark and photo-conductivity decreased with the Ge concentration because of the reduced crystalline volume fraction of the Si atoms ($X_{Si\text{-}Si}$) and the increased Ge dangling bond density. However, under higher hydrogen dilution condition, crystallization of the Si atoms decreased while crystallization of the Ge atoms increased with Ge incorporation, and the optoelectronic properties were strongly influenced by the defect density and crystallization of Ge atoms in films.
Keywords. Hydrogenated microcrystalline silicon-germanium; Optoelectronic properties; Ge concentration; hydrogen dilution.

1. Introduction

In recent years, μc-$Si_{1-x}Ge_x$:H has become increasingly attractive as an alternative bottom sub-cell absorber for multi-junction thin film silicon solar cells due to its narrower bandgap and higher absorption coefficient than microcrystalline silicon (μc-Si: H) [1-6]. Nevertheless, the optoelectronic quality depends strongly on the defect density and structural composition of the μc-$Si_{1-x}Ge_x$: H films is usually not satisfactory [6-9]. The crystallization of the μc-$Si_{1-x}Ge_x$: H films is considered complex due to random distribution of Si and Ge atoms during the film growth [7, 8]. Considering the effect of hydrogen dilution profiling on facilitating the crystallization and improving the properties of microcrystalline thin films [10, 11], in this paper, we managed to acquire the variation in crystallization of Si and Ge atoms of μc-SiGe: H films with Ge incorporation under different hydrogen dilution profiles.

2. Experiments

The μc-SiGe: H films were prepared by RF-PECVD from a mixture of SiH_4 and GeH_4 diluted in H_2. The hydrogen dilution ratio $R_H = [H_2]/\ \{[SiH_4] + [GeH_4] + [H_2]\}$ was varied from 98% to 99.5%. The Ge mass concentrations were varied from 8% to 93% in the μc-$Si_{1-x}Ge_x$: H films. The thickness of the films was in the range between 600 and 800 nm. The deposition pressure and total gas flow rate were fixed at 3torr and 400 sccm respectively, with a power density of 0.1 W/cm^2 and a temperature of 200

°C. X-ray diffraction (XRD) and Raman scattering techniques were employed to identify the crystalline phase in the films. Raman spectra were measured utilizing a He–Ne laser (λ=6328 Å) light source (LabRAM HR800). The crystalline volume fraction of silicon atoms ($X_{Si\text{-}Si}$) was calculated as $X_c=(I_c+I_i)/(I_c+I_i+I_a)$, where I_a, I_i and I_c corresponded to the integrated intensities of the Raman peaks near 480, 500 and 520 cm^{-1}, respectively [12]. Fourier transform infrared (FTIR) spectroscopy with reference to an undeposited c-Si wafer was applied to determine the molecular bonding in the μc-SiGe: H films. The dark and photo (AM1.5) conductivities were evaluated with a coplanar contact configuration. The alloy composition of μc-SiGe: H thin films was determined by X-ray fluoroscopy (XRF. NanoScope IV).

3. Results and discussion

3.1. Photoelectric properties

Figure 1 displays the photoelectric properties of μc-$Si_{1-x}Ge_x$: H thin films prepared under various hydrogen dilution ratio as a function of Ge concentration. In figure1 (a), (b) and (c), the electrical conductivity and the photosensitivity of μc-$Si_{1-x}Ge_x$: H films with R_H =98.3%, 98.5%, 98.7% and 99% are summarized as a function of Ge content x. The photoconductivity decreases between 10^{-3} and 10^{-5} S cm^{-1} for x < 50% while it increases for Ge rich samples of each series (x > 50%), and which shifts to higher Ge concentration if the hydrogen dilution is enhanced. In figure 1(b), the dark conductivity for each series all decrease monotonously by more than four orders of magnitude with x increases from 5% to 65%, and for a fixed Ge concentration the dark conductivity increases slightly with increasing R_H. The variations of the dark and photoconductivity versus hydrogen dilution for a fixed Ge fraction (x = 5~65%) are agreed with the Krause's results (Ge concentration x = 10~30%) [7].

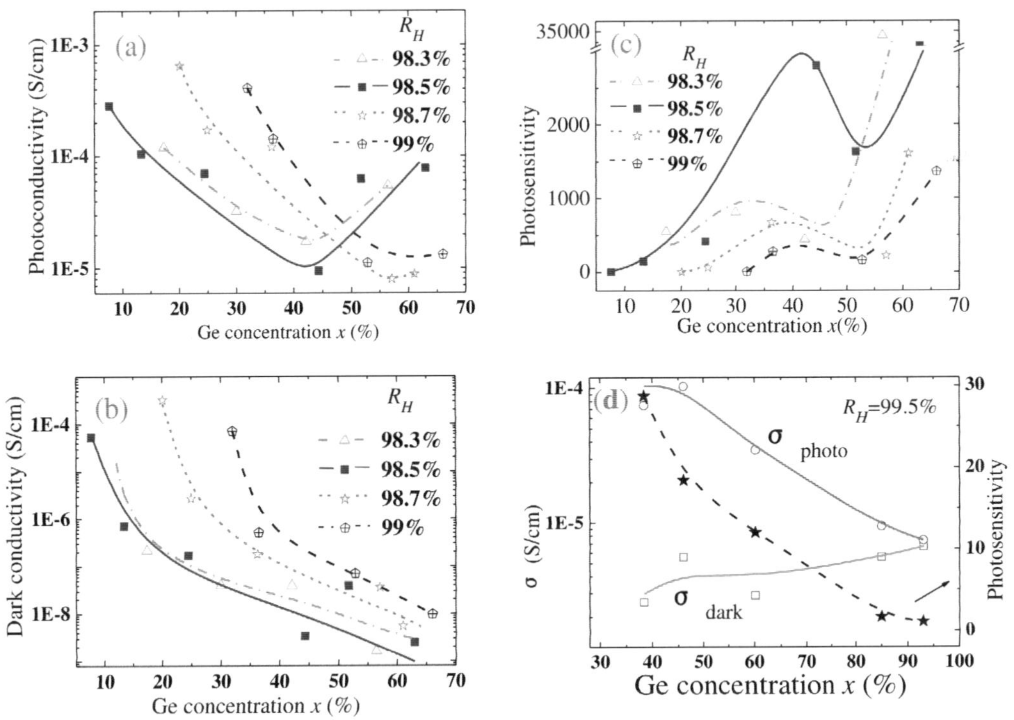

Figure 1. Photoelectric properties of μc-$Si_{1-x}Ge_x$: H thin films with various R_H as a function of Ge concentration. (a) Photoconductivity, (b) dark conductivity and (c) photosensitivity of films with R_H= 98.3%, 98.5%, 98.7% and 99%. (d) Photoelectric properties of films with R_H= 99.5%.

For the series, the photosensitivity of films increases at the beginning, and then decreases gradually with Ge content for x< 50%. A substantial increase in photosensitivity is possibly due to a transition from microcrystalline to amorphous growth for x > 50%. The photosensitivity for each series has a maximum for x <50% and it is the best when R_H=98.5%. In figure 1 (d), we further increased the hydrogen dilution ratio up to 99.5%. Different from the previous results shown in the low hydrogen dilution, as Ge concentration increases, the dark conductivity shows a small change between 10^{-6} and 10^{-5} S cm^{-1} while the photoconductivity decreases monotonously from 10^{-4} to 10^{-5} S cm^{-1}, and the photosensitivity shows a monotonic decrease from 30 and 1.

3.2. Structural characteristics of μc-$Si_{1-x}Ge_x$: H

Since the structure of the films was proved to be mainly responsible for the electrical and optical properties [10], we have investigated the structural characteristics of μc-$Si_{1-x}Ge_x$: H films prepared under various hydrogen dilution with different Ge content.

First, we investigated the crystallization of μc-$Si_{1-x}Ge_x$: H films prepared under the relatively low hydrogen dilution as a function of Ge content. Raman spectra of samples prepared under R_H = 98.7% is shown in the insert of figure 2 as a function of Ge content. For the samples with x < 24%, a peak is observed around 500 cm^{-1}; whereas for the samples with x =36%, 56%, three peaks are observed due to a high Ge concentration, which are attributed to Si-Si (~500 cm^{-1}), Si–Ge (~400 cm^{-1}) and Ge–Ge(~300 cm^{-1}) optical vibration modes [13]; for x = 61%, only Si–Ge and Ge–Ge peaks are observed. Moreover, Ge-Ge vibration peaks exhibits the characteristic spectra signals of the amorphous phases (~270 cm^{-1}) [13], which indicates that the Ge atoms are predominantly incorporated in amorphous matrix. It should be noted that similar results are obtained for the series of samples prepared under R_H = 98.3%, 98.5% and 99%. Hence, we only plotted the crystalline volume fraction of the Si atoms of films prepared under R_H = 98.3%, 98.5%, 98.7% and 99% as shown in figure 2, where the values of X_{Si-Si} decreases with the Ge concentration and drops to 0 for Ge rich samples, which is mainly attributed to an increase in the structural disorder of samples with Ge incorporation. To sum up, Raman spectroscopy for each series shows a transition from microcrystalline to amorphous growth in Si atoms with increasing Ge concentration, and Ge atoms are predominantly incorporated in amorphous matrix under the relatively low hydrogen dilution condition.

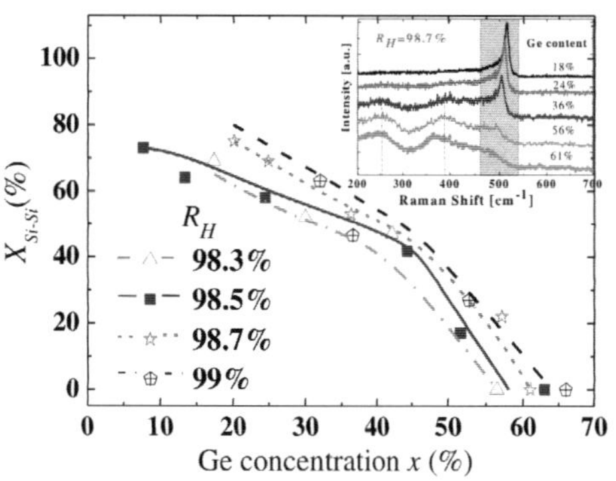

Figure 2. Crystalline volume fraction of the Si atoms (X_{Si-Si}) of samples with R_H =98.3%, 98.5%, 98.7% and 99% as a function of the Ge concentration. Raman spectra of samples with R_H = 98.7% is plotted in the insert.

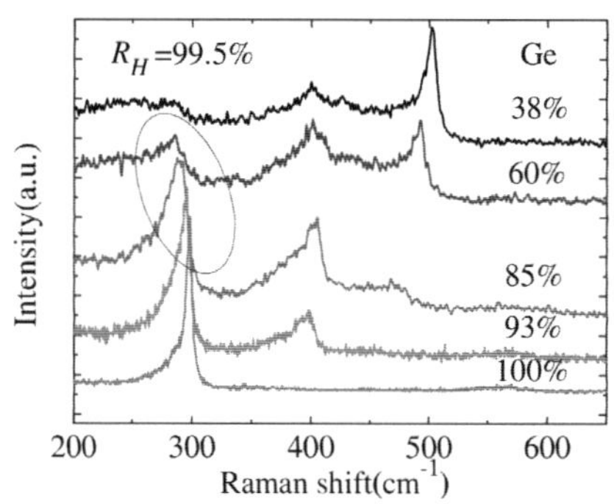

Figure 3. Raman spectra of samples with R_H = 99.5% and different Ge concentrations

Many groups have reported that hydrogen dilution is one of the key parameters for changing the structure of SiGe alloys, and the relatively higher hydrogen dilution facilitates the crystallization and improves the properties of μc-SiGe: H films [10, 11].

To obtain the dependence of the optoelectronic properties of the μc-SiGe: H films on both the crystallization of Ge atoms and Ge concentration, we further investigated Raman spectra of μc-$Si_{1-x}Ge_x$: H samples with different Ge concentrations under R_H =99.5% condition as shown in figure 3. It could be observed that crystallization of the Si atoms decreased with the Ge concentration and this behavior was very similar to that shown in low hydrogen dilution condition; while crystallization of the Ge atoms increased with the Ge content. The corresponding XRD spectra of the samples with various Ge concentrations are given in figure 4, where three peaks appears in each spectrum associated with the <111>, <220> and <311> planes of μc-SiGe:H alloys, respectively. As Ge concentration increases, the intensity of the (220) peak of samples increases, indicating that the crystallization of the films increases with the Ge concentration under the relatively high hydrogen-dilution condition. The results revealing that the material equilibrium around the Ge atoms gain exceed the losses with increasing Ge concentration when hydrogen dilution is high enough.

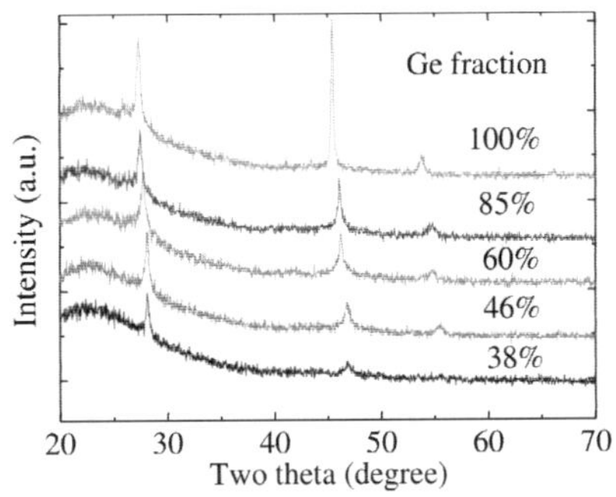

Figure 4. XRD spectra of the samples with R_H =99.5% and various Ge content x.

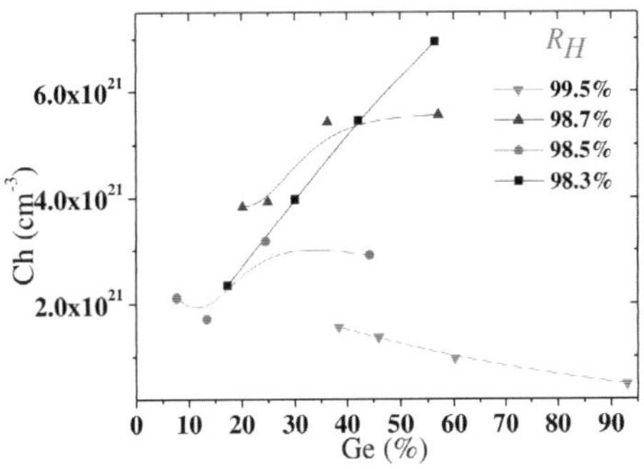

Figure 5. Hydrogen bond concentration C_H from FTIR with Ge content under R_H =98.3%, 98.5% , and R_H =99.5%

In order to verify the above result, we respectively calculated Hydrogen bond concentration (C_H) from FTIR with Ge concentration under the relatively low hydrogen (R_H =98.3%, 98.5% and 98.7%) and higher (R_H =99.5%) hydrogen dilution condition as shown in figure 5. It could be observed that C_H increased with x under R_H =98.3%, 98.5% and 98.7% due to the increase of Ge dangling bonds, while C_H decreased with x under R_H =99.5%, revealing that the increased crystallization of Ge atoms played a role in the decreasing of C_H under the relatively high hydrogen-dilution condition.

We correlated the above result with optoelectronic properties. When the μc-$Si_{1-x}Ge_x$: H films are prepared under the relatively low hydrogen dilution condition, the decrease in photoconductivity (Fig. 1 (a)) can be attributed to the reduced minority carrier lifetime associated with the increased Ge dangling bond density and the decreased crystallization of the Si atom, which led to a monotonic decrease in the dark conductivity of μc-$Si_{1-x}Ge_x$: H with Ge incorporation as illustrated in figure 1 (b). Moreover, for Ge rich films (x>50%), the photoconductivity of μc-$Si_{1-x}Ge_x$: H films increased mainly due to an electrical transition from weak n-type to strong p-type conduction according to Refs, 6 and 9. It should be underlined that the photosensitivity (figure 1 (c)) of μc-$Si_{1-x}Ge_x$: H did not show a monotonic change with crystallization of the Si atoms ($X_{Si\text{-}Si}$), which is different from that of μc-Si:H [14]. Nevertheless, we considered that the increased defect density and the excessive grain size caused by the increased crystallization of Ge atoms were mainly responsible for the deterioration in photoelectric properties of μc-$Si_{1-x}Ge_x$: H films under the relatively high hydrogen dilution.

4. Conclusions

Effect of alloy composition on the optoelectronic properties of the μc-$Si_{1-x}Ge_x$: H films was studied under various hydrogen dilution conditions. The optoelectronic properties were strongly influenced by Ge concentration and crystal structure of films. Under the relatively low hydrogen dilution condition

(R_H =98.3~99%), Ge atoms are predominantly incorporated in an amorphous matrix and the photosensitivity has a maximum when the crystallization of Si atoms is in the range of 45-55%. However, when hydrogen dilution is high enough (R_H =99.5%), the increased crystallization of Ge atoms in high Ge content (x > 0.5) μc-$Si_{1-x}Ge_x$: H films plays a more important role in the optoelectronic properties.

5. References

[1] Cao Y, Liu Y M, Zhou J, Wang Y J, Ni J and Zhang J J, 2016 Non-uniform distribution in μc-$Si_{1-x}Ge_x$: H and its influence on thin film and device performance, *Sol. Energy Mater. Sol. Cells* **151** 1.

[2] Bidiville A, Matsui T and Kondo M 2014 Effect of oxygen doping in microcrystalline SiGe p-i-n solar cells, *J. Appl. Phys.* **116** 053701

[3] Ni J, Liu Q, Zhang J J, Ma J, Wang H, Zhang X D and Zhao Y 2014 Microcrystalline silicon–germanium solar cells with spectral sensitivities extending into 1300 nm, *Sol. Energy Mater. Sol. Cells* **126** 6

[4] Cao Y, Zhang J, Li C, Li T, Huang Z, Ni J, Hu Z, Geng X and Zhao Y, 2013 Hydrogenated microcrystalline silicon germanium as bottom sub-cell absorber for triple junction solar cell, *Sol. Energy Mater. Sol. Cells* **114** 161.

[5] Li T W , Zhang J J, Ma Y, Yu Y W and Zhao Y, 2017 Effect of crystal structure of Ge atoms on optoelectronic properties of hydrogenated microcrystalline Si-Ge films, *Modern Physics Letters B*, **31** 1740010.

[6] Matsui T, Kondo M, Ogata K, Ozawa T and Isomura M, 2006 Influence of alloy composition on carrier transport and solar cell properties of hydrogenated microcrystalline silicon–germanium thin films, *Appl. Phys. Lett.* **89** 142115.

[7] Krause M, Stiebig H, Carius R, Zastrow U, Bay H and Wagneret H, 2002 Structural and optoelectronic properties of microcrystalline silicon germanium, *J. Non-Cryst. Solids* **299–302** 158.

[8] Rath J K, Tichelaar F D, Ruud E and Schropp I, 2002 Heterogeneous growth of microcrystalline silicon germanium, *Sol. Energy Mater. Sol. Cells* **74** 553.

[9] Matsui T, Ogata K, Chang C W, Isomura M and Kondo M, 2008 Carrier collection characteristics of microcrystalline silicon–germanium p–i–n junction solar cells, *J. Non-Cryst. Solids* **354** 2468.

[10] Vetterl O, Finger F, Carius R, Hapke P, Houben L, Kluth O, Lambertz A, Mück A, Rech B and Wagner H, 2000 Intrinsic microcrystalline silicon: A new material for photovoltaics, *Solar Energy Mater.* **62** 97.

[11] Krause M, Carius R, Stiebig H and Wagner H, 2000 *Mater. Res. Soc. Symp. Proc.* **609** A13.7.

[12] Tsu R, Gonzaicz J, Chao S S, Lee S C and Tanaka K, 1982Critical volume fraction of crystallinity for conductivity percolation in phosphorus-doped Si:F:H alloys, *Appl. Phys. Lett.* **40** 534.

[13] Renucci M A, Renucci J B and Cardona M, 1971 in: M.Balkan-ski (Ed.), Light Scattering in Solids, Flammarion, Paris, 326.

[14] Ballif C, Beyer W, Bouree J E, Droz C, Finger F, Python M, Randall J, Schade H and Wyrsch N, in: A. Shah (Ed.), 2010, *Thin-film silicon solar cells*, EPFL Press 127.

Acknowledgments

This work was supported by the National Science Foundation of Liaoning Province (Grant No. 20180510017, Grant No. 20170540768).

IOP Conf. Series: Materials Science and Engineering **761** (2020) 012006 doi:10.1088/1757-899X/761/1/012006

Biocompatibility, Morphology, and Chemical Elements Composition of Indigenous Indonesian Wild Silkmoth Cocoon for Future Application in the Field of Biomaterial

T. G. T. Nindhia[1]*, Z. Knejzlik[3], T. S. Nindhia[2], I W. Surata[1], Tomas Ruml[3]
[1] Department of Mechanical Engineering, Engineering Faculty Udayana University, Jimbaran, Bali, Indonesia.
*E-mail: nindhia@yahoo.com
[2] Faculty of Veterinary Medicine, Udayana University, Sudirman Campus, Denpasar, Bali, Indonesia
[3] Faculty of Food and Biochemical Technology, Technicka 5, 166 28 Prague, Czech Republic

Abstract. Morphology of two indigenous Indonesian wild silkworm cocoons from the species of *Cricula trifenestrata* and *Attacus atlas* were identified and compared with common silkworm cocoon of *Bombyx mori* that was also grown in Indonesia. The equipment that was utilized for identification processes was scanning electron microscope (SEM). The Laser Induced Breakdown Spectroscopy (LIBs) was used to identify the elements composition in the respective silkworm cocoon. The cell from the Type of U-2 OS was used for test of biocompatibility of the silkworm cocoon under cell culturing. The morphologies are presented in the report and the chemical element compositions are reported respectively. It is Interesting to be explored that both Indonesian wild silkworm cocoon have good biocompatibility as the domesticated silkworm *Bombyx mori*.

1. Introduction

Silks are material that mechanically superior, biocompatible which is also degradable. Silk; therefore offer a wide range of properties [1]. The mechanical properties of silks fiber in terms of energy absorbed before failure exceed all other natural polymers and synthetic materials, even high-performance fibers such as Kevlar [2].

As reported in many previous publications, the silk from the species of *Bombyx mori* is well known for its biocompatibility as biomaterial. The domesticated *Bombyx mori* silk has been used commercially as sutures in biomedical [3].

Silk then have more attention as biomaterial since many researcher recently was demonstrated that silk is suitable for 3-D scaffolding material. As a scaffold, the cell growth and spread along the fiber, and the next, covered all the surface of the fiber and then spread to fill the gap to form structure of the tissue [4]. It was also reported that the human cell was adhere, grow and spread on three dimensional silk fibroin produced from cocoons of *Bombyx mori* [5].

Utilization of silkworm cocoon as biomaterial as above mentioned until now just put attention on one species of silkworm cocoon that is form the species of *Bombyx mori*. Other type of silkworm cocoon, especially from the wild that is not domesticated yet should be explored in other to find more suitable silk material for the better or even more superior biocompatibility. It is the purpose of this research to observe the Indonesian source of wild silkworm cocoons from the species of *Cricula trifenestrata* and *Attacus atlas* to be investigated its morphology, chemical elements composition and biocompatibility.

2. Experimental

The wild Silkworm cocoon form the species of Cricula *trifenestrata* (Figure 1a) and *Attacus atlas* (Figure 1b) were obtained from Indonesian sources. The domesticated silkworm cocoon of *Bombyx mori* that was also from Indonesian source (Figure 1c) was also included in this research for the comparison as a standard material.

The observation on morphology was conducted by using scanning electron microscope (SEM). The photo was taken and the diameter was measured. It was selected 100 single fibers from the cocoon randomly and the result was presented in histogram for easy comparison.

(a) (b) (c)

Figure 1. Cocoon of *Cricula trifenestrata* (a), *Attacus atlas* (b), and Bombyx *mori* (c)

The laser induced breakdown spectroscopy (LIBs) was used to identify the element content in the respective silkworm cocoon. Due to limitation of the space for this article, only the element from the wave length range from 0 to 500 nm was presented.

The biocompatibility test was conducted as follows: unwoven fibers were sterilized with ethanol for 2 days at room temperature. Samples were washed by PBS. Suspension of human osteosarcoma cell line in Dulbecco's modified Eagle's cultivation medium was prepared and then the fibers were soaked in this medium at atmosphere with 5% CO_2. The medium was supplemented with 10% FBS. The cells were cultivated for 2 days. For observation of the cell, the sample was taken out from cultivation medium and transferred to PBS and washed. Fluorescence microscope was utilized to observe the cell growth as characteristic of biocompatibility.

3. Result and discussion

The morphology of the fiber from the *Bombyx mori* is presented in Figure 2a. It is found by measurement that the fiber diameter is 30% appear in frequency of the histogram for the size 31-32 µm (Figure 2). This is mean that the fiber size is smaller that both *Cricula trifenestrata* and Attacus *atlas*. The *Cricula trifenestrata* is found with diameter 55-56µm or 58-59µm (appears the same 20% in frequency of the histogram) as depicted in Figure 3a and Figure 3b. The diameter of the *Attacus attlas* is seem in between of *Bombyx mori* and *Cricula trifenestrata* that is 40-41 µm (appear 30% in frequency of the histogram in Figure 4)

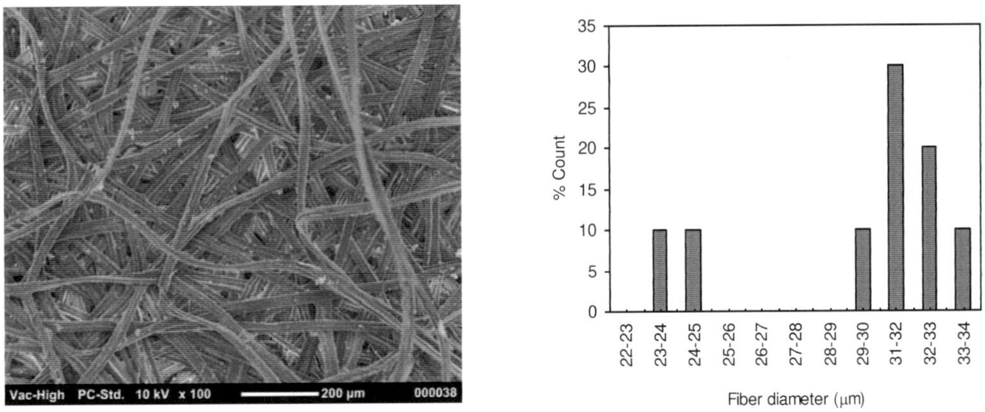

(a) (b)

Figure 2. (a) Marphology of *Bombyx mori* (b) Histogram of the diameter of the fiber of *Bombyx mori*

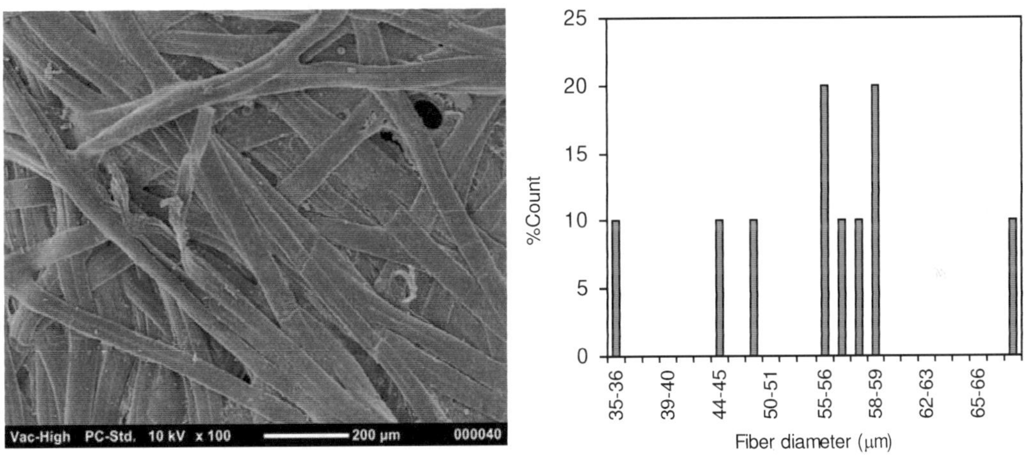

(a) (b)

Figure 3. (a) *Marphology of Cricula trifenestrata*, (b) Histogram of the diameter of the fiber of Cricula

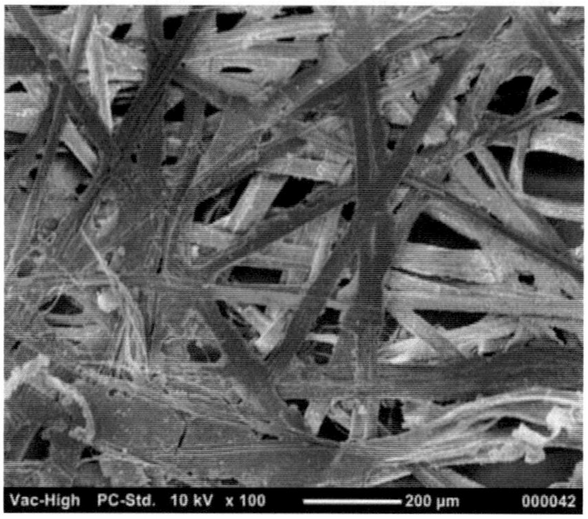

(a)

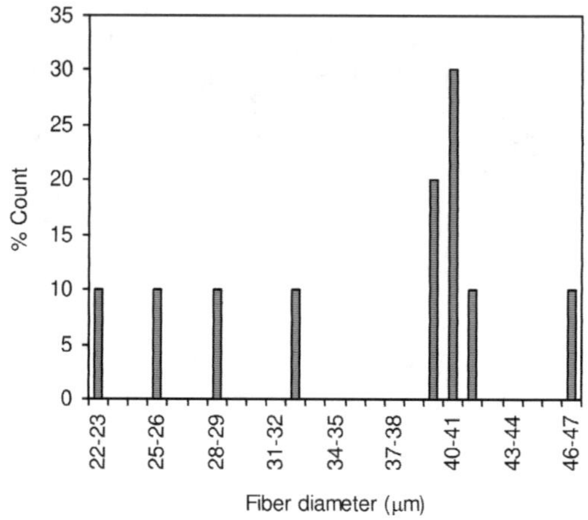

(b)

Figure 4. (a) Morphology of *Attacus atlas* (b) Histogram of the diameter of the fiber of *Attacus atlas*

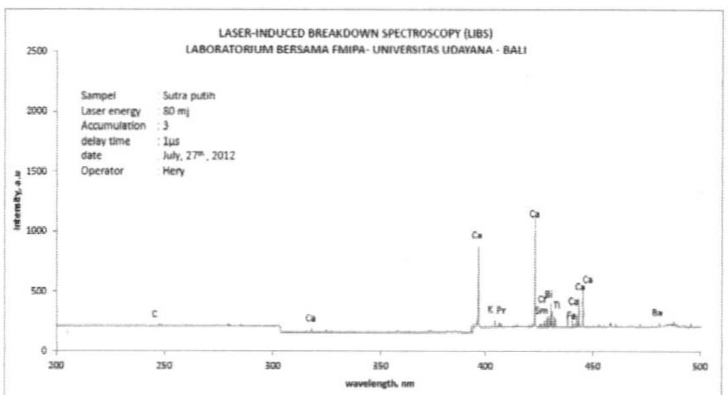

(a)

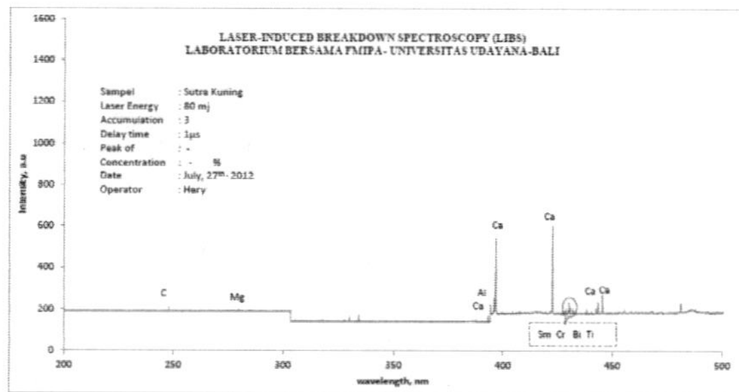

(b)

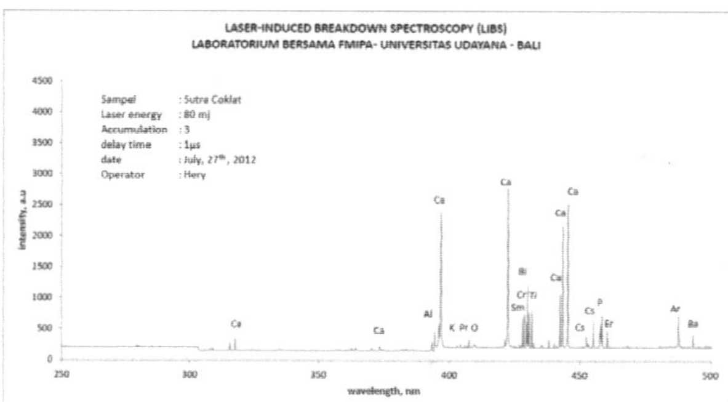

(c)

Figure 5. Element analyses result by using laser induced breakdown spectroscopy (LIBs) (a) *Bombyx mori* (b) *Cricula trifenestrata* (c) *Attacus atlas*

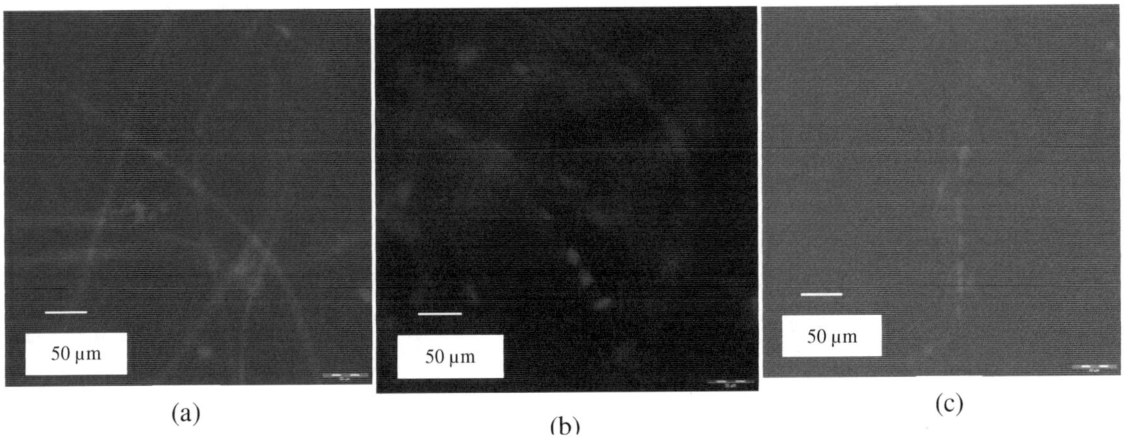

Figure 6. Growth of the cells (a) Bombyx *mori* (b) *Cricula trifenestrata* (c) *Attacus atlas*

It can be proofed by the laser induced breakdown spectroscopy (LIBs) that the chemical element content each cocoon are found different as depicted in Figure 5. The chemical element content in the *Bombyx mori* are: C, Ca, K, Pr, Sm, Cr, Bi, Ti, Fe, Ba. Chemical elements content for the *Cricula trifenestrata* are: C, Mg, Al, Sm, Cr, Bi, Ti. There are about 14 elements that are found in the *Attacus atlas* namely: Ca, Al, K, Pr, O, Sm, Cr, Bi, Ti, Cs, P, Er, Af, Ba.

It is interesting to discuses here related the biocompatibility of the wild silkworm cocoons of Indonesian source. It was explained in the previous publication [6, 7, and 8] that *Bombyx mori* is biocompatible. In this research is also can be proofed that the *bombyx mori* is biocompatible. Figure 6 is the microscope observation of the growth of the cell in the respective silkworm cocoon for 2 days. It is clearly observed that the cell are found more in the both of *Cricula trifenestrata* and *Attacus atlas* (Figure 6b and 6c) comparing in the *Bombyx mori*. This condition can be directed to the result that wild silkworm cocoon from Indonesia are more biocompatible comparing the *Bombyx mori* that usually use in recent day for biomaterial research.

Researcher still put attention to the strength of biomaterial and put attention mainly for alloy as solution for implant [9]. The modulus elasticity of implant alloys are still far from modulus elasticity of living organs and still not appropriate in biocompatibility properties. Using natural materials such a silk as biomaterial for organs replacement are promising since it has much closer modulus elasticity to living organ and also has better biocompatibility.

4. Conclusion

The morphology of the Silkworm cocoon of *Cricula trifenestrata* and *Attacus atlas* were found different with *Bombyx mori*. The diameter of *Bombyx mori* fiber is about half of the *Cricula trifenestrata* or *Attacus atlas*. The chemical element composition are proofed different but all are found having good biocompatibility for cell culturing with note that it seem both Indonesia wild silkworm cocoon having better biocompatibility than *Bombyx mori*.

5. References

[1] Veparia, C., Kaplan, D.L., Silk as a biomaterial, Prog. *Polym. Sci.*, vol. 32, 2007, pp. 991–1007.
[2] Meinel, L., Fajardod, R., Hofmanna, S., Langerc, R., Chene, J., Snyderd, B., Novakovic, G.B, . Kaplan, DL., Silk implants for the healing of critical size bone defects, *Bone*, vol. 37, 2005, pp. 688–698.
[3] Altman, G.H., Diaz, F., Jakuba, C., Calabro, T., Horan, R.L., Chen, J., Lu, H.,, Richmond, J., and Kaplan, D.L., Silk-based biomaterials, *Biomaterials*, vol. 24, 2003, pp. 401–416.

[4] R. E. Unger , K. Petersa, M. Wolf, A. Motta, C. Migliaresi, C. J. Kirkpatrick, Endothelialization of a non-woven silk fibroin net for use in tissue engineering: growth and gene regulation of human endothelial cells, *Biomaterials*, vol. 25, 2004, pp. 5137–5146.
[5] Unger, R.E, Wolf, M., Peters, K., Motta, A., Migliaresi, C., Kirkpatrick, C.J., Growth of human cells on a non-woven silk fibroin net: a potential for use in tissue engineering, *Biomaterials*, vol. 25, 2004, pp. 1069–1075.
[6] Nindhia, T.S., Nindhia, T.G.T, Surata, I W., Knejzlik, Z., Ruml, T., effect of feeding with herb of *erythrina variegata* to biocompatibility of the cocoon fiber of wild silk moth *attacus atlas* for future application as biocompatible of silk sutures, *Asian J Pharm Clin Res*, Vol 11, Special issue 3, 2018, pp. 20-23.
[7] Nindhia, T.G.T., Knejzlik, Z., Ruml, T., Surata, I W. and Nindhia, T.S., Indigenous Indonesian Wild Silkworm Cocoon of Attacus atlas as Biocompatible Film Biomaterial, *Materials Science and Engineering* 204, 012011, 2017, pp.1-5.
[8] Nindhia, T.G.T., Koyoshib, Y., Kanekob, A., Sawada, H, Ohtab, M., Hiraib, S., Uoc, M., Hydroxyapatite-silk Functionally Graded Material by Pulse Electric Current Sintering, *Trends Biomater. Artif. Organs*, Vol 22(1), 2008, pp 28-33.
[9] Diana,I. B., New Metallic Alloys Used for Dental Implants Manufacturing, *International Journal of Materials, Mechanics and Manufacturing,* Vol. 6, No. 2, 2018, pp 142-146.

Acknowledgment
The authors acknowledge financial support received from *Kementrian Riset Teknologi dan Perguruan Tinggi* through Udayana University for the year 2019 under contract number: 492.18/UN14.4.A/LT/2019. Partial support from the Institute of Chemical Technology of Prague also very much acknowledged. Special thank addressed to Prof. Ing. Katerina Demnerova, CSc. Together with Ing. Jan Lipov, Ph.D For the permisison to use all research facility in the Department of Biochemical and Microbiology Institute techmology of Prague, Czeck Republic. This work also got funding for international conference participation (*Bantuan Seminar Luar Negeri/BSL*) from Ministry of research, Technology and Higher Education the Republic of Indonesia under Letter No. B/2994/E5.3/KI.03.01/209 dated 24th September 2019.

Assessment of Antifungal Activity of Aloe Vera Toothpaste against Candida Albicans

S Thaweboon [1] and B Thaweboon [2]
[1] Professor, Department of Oral Microbiology, Faculty of Dentistry
[2] Associate Professor, Department of Oral Microbiology, Faculty of Dentistry
Mahidol University, Bangkok 10400, Thailand

E-mail: sroisiri.tha@mahidol.ac.th

Abstract. Herbal products have been the basis for medical therapy and are noted as a form of alternative medicine. Modern medicine also employs the use of many herbal products for therapeutic purposes. Aloe vera is a medicinal herb whose mucilaginous gel is traditionally used to treat different diseases. It consists of the active compounds which include methylchromones, flavonoids, aloesin, aloe-emodin, sterols, amino acids, aloemannan, aloin, acemannan, aloeride, naftoquinones, saponin and vitamins. Aloe vera toothpaste, consequently, has been developed by the Thailand Institute of Scientific and Technological Research (TISTR) to fulfill a strategy of good oral hygiene. The aim of this study was to investigate the antifungal activity of aloe vera toothpaste against *Candida albicans* in vitro. Toothpaste containing 10% aqueous extract of the aloe vera leaf was prepared by TISTR. The toothpaste solution was prepared by mixing 0.66 g of toothpaste in 1 mL of sterile deionizing water. The suspensions of 107cfu/mL (equivalent to 0.5 McFarland Standard) of *Candida albicans* ATCC 10231 and 2 clinical isolates were prepared and added to toothpaste solution. After 50 sec, the final solution was transferred to D/E neutralizing broth and cultured at 37 °C for 24-48 h on Sabouraud dextrose agar. Sterile distilled water was used as a control. Colony forming unit (CFU) was counted after incubation. The result showed that aloe vera toothpaste had a significant inhibitory effect against tested candida species compared with a control. In conclusion, result from the present study is scientific information to prove that aloe vera toothpaste could be assigned as natural toothpaste against candida infection. With proper tooth brushing, it is a simple manner, leading to good oral care.

1. Introduction

An overgrowth of Candida in the oral cavity can cause candidosis, which is a common opportunistic infection. The disease is frequently seen in infants and older adults, and also with states of local and systemic immunological suppression such as human immunodeficiency virus (HIV) infection, leukemia, diabetes, systemic chemotherapy, radiation therapy, malnutrition, and the use of systemic corticosteroids and broad-spectrum antimicrobials. Apart from candidosis, Candida species are also indicated to be caries-associated microorganism and are generally used as indicators of caries activity. The high salivary candida carriage has been associated with the severity of caries among children and adults and the most often found are *C. albicans* [1-4].

Aloe vera is a medicinal herb whose mucilaginous gel is traditionally used to treat different diseases. It consists of the active compounds which include methylchromones, flavonoids, aloesin, aloe-emodin, sterols, amino acids, aloemannan, aloin, acemannan, aloeride, naftoquinones, saponin and vitamins [5].

In dentistry, herbal toothpastes containing aloe vera extract were shown to have significant antimicrobial activity against *Streptococcus* species and *Actinomyces* species as well as an improvement in gingival and plaque index scores of gingivitis patients [6]. In vitro study by Bernardes *et al* [7] demonstrated that aloe vera extract (10% v/v) reduced both growth and germ tube formation by C. *albicans*. In addition, antiplaque and antifungal effectiveness of aloe vera toothpaste was reported among a group of intellectually disabled adolescents [8]. Therefore, the objective of this study was to investigate the antifungal activities of toothpaste containing aloe vera extract against *C. albicans*.

2. Materials and methods

2.1. Toothpaste
Toothpaste containing 10% aqueous extract of the aloe vera leaf was prepared by the Thailand Institute of Scientific and Technological Research (TISTR), Ministry of Science and Technology, Thailand. Toothpaste solution was prepared for the use in further experiment by mixing 0.66 g of toothpaste in 1 mL of sterile deionizing water.

2.2. Microorganisms
Pure cultures of *Candida albicans* ATCC 10231 and 2 strains isolated from patients' oral lesions were obtained from the culture collection of Oral Microbiology Department, Faculty of Dentistry, Mahidol University. They were maintained in freeze drying condition (lyophilization). Afterwards, they were subcultured on Sabouraud dextrose agar (Difco Laboratories, Detriot, MI) and incubated at 37°C for 24 h. The colonies were inoculated in BHI broth and suspension of 107cfu/mL (equivalent to 0.5 McFarland Standard) of each strain was prepared for antifungal determination.

2.3 Antimicrobial assay
One mL of the prepared candida suspension was added to the toothpaste solution to make a final concentration of 22%, which was equal to the initial concentration of toothpaste in normal tooth brushing [9]. After 50 sec, 10 μl of the solution was transferred to 190 μl of D/E neutralizing broth (Difco Laboratories, Detriot, MI), further 10-fold diluted and cultured at 37 °C for 24-48 h on Sabouraud dextrose agar for the quantitation of *C. albicans*. Sterile distilled water was used as a control. Colony forming unit (CFU) grown on agar plate was counted after incubation.

2.4 Statistical analysis
A statistical analysis was made by Mann-Whitney tests to compare the effect of toothpaste with a control. A p-value of <0.05 was noticed significant. For illustration, the results are presented as means and standard deviations of Log CFU/mL.
All tests were carried out in triplicate on three separate occasions.

3. Results
The results of this investigation showed that aloe vera toothpaste had a significant suppressive effect against all tested candida species compared to control (Table 1).

IOP Conf. Series: Materials Science and Engineering **761** (2020) 012007 doi:10.1088/1757-899X/761/1/012007

Table 1. The number of C.albicans after exposure to toothpaste solution

Toothpaste	*C. albicans*		
	ATCC 10231	clinical isolate I	clinical isolate II
Control	5.03±0.15	5.18±0.12	5.75±0.06
Aloe vera	1.63±0.06	2.24±1.01	4.01±0.15

* data expressed as mean±SD of Log colony forming unit (CFU)/mL

4. Discussion

Several chemical agents are commercially available for the treatment of infectious diseases; however, they can shift the balance of oral microbial ecology, cause resistant strains and have undesirable adverse effects including hypersensitivity, immunosuppressive and allergic reactions. As a result, the search for new substances with antimicrobial properties to be used against these microorganisms is necessary. Plant products or natural products have been demonstrated to have an important role in diseases prevention and treatment through the enhancement of antioxidant activity, inhibition of microbial growth, and modulation of genetic pathways. Various natural components or phytochemicals isolated from plants are considered as good alternatives. In addition, the World Health Organization (WHO) has estimated that about 80% of the population still depends upon herbal medicines for the treatment of several diseases as a result of simple availability, economic reasons, and less side effects [10].

In dentistry, the aloe vera gel has been tried for healing of aphthous ulcers, alveolar osteitis, and lichen planus lesions. This gel has also been found to have strong bactericidal activity against some cariogenic (*Streptococcus mutans* and *Streptococcus mitis*) and periodontopathic (*Aggregatibacter actinomycetemcomitans, Porphyromonas gingivalis,Prevotella intermedia, Peptostreptococcus anaerobius* and *Bacteroides fragilis*) bacteria [11].

In this study, we developed toothpaste containing aqueous extract of aloe vera leaf and tested against 3 strains of *C. albicans*. All tested strains were incubated with 0.33 g/mL (final concentration) of toothpaste which is equivalent to 22% of the initial toothpaste concentration found in the oral cavity during normal tooth brushing. The result shows the significant suppressive effect of aloe vera toothpaste against all tested strains. The most effective result was demonstrated on the reference strain of *C. albicans* (ATCC 10231). Significant variation was reported on candida activity (e.g. biofilm mass, growth activity) between clinical and laboratory reference strains. The clinical strains were shown to have less biofilm mass and growth activity which has been suggested as a mechanism of antimicrobial resistance [12]. Similar to a previous study, 0.5 – 10% aqueous extract of aloe vera leaf was demonstrated to produce a severe decrease in *C. albicans* growth and the ability to form a germ tube [13]. However, the susceptibility of candida to antimicrobial agents, especially those extracted from plants, may vary due to the geographical and climatic conditions which will affect the phytochemical composition of the plants and its antimicrobial activity.

The components of anthraquinones in aloe vera are thought to be responsible for its antimicrobial activity. There are complex and more than 20 kinds of substances, such as aloin, aloesin and aloe-emodin. The antimicrobial mechanisms of anthraquinonein might be 1) the change in cell wall synthesis and cell membrane permeability which leads to cell damage (by interfering absorption of nutrients from the outside and the leakage of nucleic acids causes the cell deformation), 2) inhibiting the respiratory metabolism (by inhibiting the Krebs cycle of oxidation of sugar and the synthesis of

ATP and NADH in the cell), 3) inhibition of nucleic acid synthesis (by its ability to bind with phosphate group of DNA which affect the replication, transcription and lead to cell death) [14-16].

5. Summary
Toothpaste containing aloe vera extract was demonstrated to have significant inhibitory effect against *C. albicans*. Data from this study could be scientific evidence to demonstrate that aloe vera toothpaste could be employed as natural antifungal toothpaste. With proper tooth brushing, it is a simple manner, leading to good oral care.

6. References
[1] Rajam M, Hannan A, Ali K 2010 *Caries. Res.* vol 44 p 272
[2] Xiao J, Huang X, Alkhers N, et al 2018 *Caries. Res.* vol 52 p 102
[3] Lozano Moraga CP, Rodríguez Martínez GA, Lefimil Puente CA, Morales Bozo IC, Urzua Orallana BR 2017 *Acta. Odontol. Scand.* vol 75 p 30
[4] De-la-Torre J, Marichalar-Mendia X, Varona-Barquin A, *et al* 2016 *Mycoses* vol 59 p 234
[5] Moghaddasi S, Verma SK 2011 *Int. J. Bio. Med. Res.* vol 2 p 466
[6] Pradeep AR, Agarwal E, Naik SB 2012 *J. Periodontol.* vol 83 p 797
[7] Bernardes I, Rodrigues MP, Bacelli GK, *et al* 2012 *Mycoses* vol 55 p 257
[8] Khatri SG, Samuel SR, Acharya S, Patil ST 2017 *Pediatr. Dent.* vol 39 p 434
[9] Duke SA, Forward GC 1982 *Br. Dent. J.* vol 152 p 52
[10] WHO global report on traditional and complementary medicine. 2019
[11] Fani M, Kohanteb J 2012 *J. Oral. Sci.* vol 54 p 15
[12] Alnuaimi AD, O'Brien-Simpson NM, Reynolds EC, McCullough MJ 2013 *FEMS. Yeast. Res.* i.vol 13 p 689
[13] Shireen F, Manipal S, Prabu D 2015 *SRM. J. Res. Dent. Sci.* vol 6 p 92
[14] Chunxia LU, Hongxin W, Wenping LV, *et al* 2011 *Fisheries. Science.* vol 77 p 375
[15] Liu M, Peng W, Qin R, Yan Z, Cen Y 2015 *Appl. Microbio, Biotech.* vol 99 p 7699
[16] Srinivas G, Babykutty S, Sathiadevan PP, Srinivas P 2007 *Med. Res. Rev.* vol 27 p 591

Acknowledgments
The authors would like to thank the staff of Thailand Institute of Scientific and Technological Research (TISTR), Ministry of Science and Technology, Thailand for their kind preparing aloe vera toothpaste for this study.

Direct Electrodeposition of Zno Nanosheets Film on Plastic Substrate

Chuanhai Xiao, Shang Sun, Qi Lai*
College of Vanadium and Titanium, Panzhihua University, Panzhihua 617000, China

Abstract. A novel and simple method to directly assemble nanostructures onto plastic substrate via a quasi-2D electrodeposition approach was proposed. It was demonstrated that such synthetic method was feasible for flexible electronics, as exemplified by the fabrication of ZnO nanosheets film photoresponse devise onto flexible transparent, polyethylene terephthalate (PET) substrate that can withstand bending. This direct assembly strategy is general to other nanostructures onto plastic substrate which can be electrodeposited.
Keywords. Electrodeposition; Flexible; Plastic substrate; ZnO nanosheet.

1. Introduction

In the recent years, flexible electronics have got considerable research enthusiasm because of their specific features and potential applications in flexible displays, sensors, electronic textiles, artificial skins and other wearable electronics, *etc* [1-5]. The academic hotspot in flexible electronics will continue to move forward, which is driven by the increased demand for electronics permitting light-weight design, portability, and low manufacturing cost as compared to their rigid substrate counterparts, and supported by techniques for the unceasingly miniaturization of individual ingredient in microelectronics [6-8]. With unique geometry, outstanding photoelectric properties, excellent mechanical flexibility and fine transparency, low-dimensional inorganic semiconductor nanomaterials facilitate significant insights and opportunities and favourable prospect for flexible electronics.

In general, traditional flexible substrates include plastics, paper, metal foils, ultra-thin glass, *etc*. The keen interest in exploiting plastics as substrates for electronic applications is derived from their advantages of flexibility, transparency, insulation, biocompatibility, light weight, low cost, and roll-to-roll processing [9, 10]. As well known, by conventional electrochemical deposition, materials cannot be sedimentated directly onto insulating plastic substrate. Nevertheless, in ultrathin layers electrodeposition, the electrode and the substrate are disconnected, and the substrate definitely be insulating. Therefore, materials can be deposited on the plastic substrate straightforward. In previous reports about the ultrathin layers electrodeposition, the nanostructure materials were none exception deposited onto rigid substrates, such as glass or SiO_2 [11-21]. It gravely limits the application range of

IOP Conf. Series: Materials Science and Engineering **761** (2020) 012008 doi:10.1088/1757-899X/761/1/012008

the obtained nanostructured materials. The insulating polymer substrate, as opposed to the metallic substrates, enables the fabrication of electrical devices. In this communication, as an example, ZnO nanosheets film was deposited onto PET substrate. Furthermore, the electrical property under different bending condition has been implemented.

2. Experimental

2.1 Method of synthesis of ZnO nanosheets film onto PET substrate

All the chemicals were analytical grade reagents without further purification. The electrolyte solution of 0.05 M Zn $(NO_3)_2$ was prepared by Millipore water. The growth system consist of a growth chamber, a low temperature cycle water bath, a DC power supply (DF1731SB5A), an arbitrary function generator (AFG 310), an optical microscope (Leica Dmlm) and a CCD camera (A311f). The core part of growth chamber is shown in Figure 1. A Peltier element attached to DC power supply was fixed in the centre of the growth chamber, which was used to rapidly adjust the electrolyte temperature. The growth chamber temperature was controlled by low-temperature cycle water bath, and the growth potential was supplied by an arbitrary function generator. The growth process could be observed by optical microscope and CCD camera. Silicon substrates were cut into pieces about 20×20 mm^2 before they were successively cleaned with ethanol and water for 10 min, rinsed with Millipore water, and blown dry with high-purity nitrogen. Then the silicon wafer was placed on the Peltier element, afterwards two parallel electrodes made of 30 μm thick Zn foil were put on silicon substrates, and separated by a distance of approximately 8 mm. Then the electrolyte was dropped on the silicon substrate. Finally, a PET sheet was carefully put on the two electrodes, after that the electrolyte would fill the space between the PET wafer and silicon substrate. A uniform ultra-thin ice layer could be formed between the silicon and PET substrate by adjusting the temperature of the Peltier element and the low-temperature cycle water bath. After an ultra-thin ice layer was formed, we kept the growth chamber temperature constant for 30 min. The ultrathin liquid layer of concentrated electrolyte was formed between the ice layer and upper surface of silicon substrate, as well as the ice layer and lower surface of PET substrate. Electrodeposition was initiated when a constant voltage of 0.8 V was applied through the parallel electrodes. When the growth process ended, a slight increase in temperature was provided by the Peltier element to melt ice. Then the PET sheet was taken out, cleaned by the deionized water, and dried at room temperature.

The morphology, component and crystal structure of the ZnO nanosheets film were characterized by scanning electron microscope (SEM FEI Magellan-400), X-ray diffraction (Rigaku D/max-Ra) and transmission electron microscopy (TEM, JEOL JEM-2200FS).

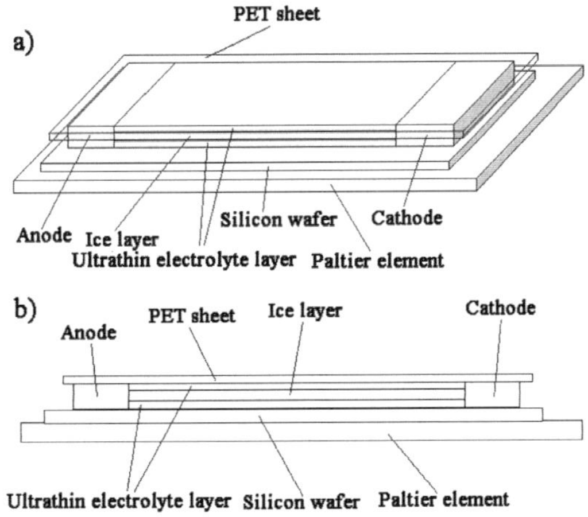

Figure 1. Schematic showing the central configuration of the electrochemical cell. (a) side- and (b) sectional view.

2.2 Electrical devices fabrication and characterization

The samples onto the PET substrate were chosen as the test object. First, the sample was covered by a linear mask with a width of 50 μm. Two copper wires were fixed on both sides of the mask separately. Then the Au film was deposited on the glass by vacuum ion sputtering for 3 min. After that, the mask was taken away, and the sample was connected into the circuit. The schematic diagram of the fabricated electrical device is shown in figure 2(a). For photocurrent measurements, the 355 nm laser was used to irradiate the device while a source voltage was applied directly to the photoconductor and the DC current was read from a Keithley 2400 source meter. For bending experiments, photocurrent measurements were carried out on the ZnO devices grown onto PET substrates that were taped on the outer surface of glass vials with various radii (r=1.3, 1.6 and 2.1 cm), as shown in figure 2(b).

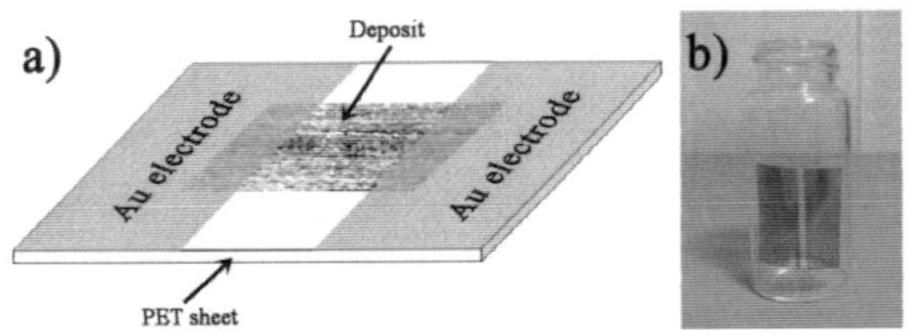

Figure 2. (a) Schematic diagram of the electrical device fabrication. (b) A picture of the fabricated flexible electrical device taped around a glass vial with a 1.3 cm radius.

3. Results and discussion

The morphology, component and structure information of ZnO nanosheets film grown onto PET substrate were investigated by SEM, XRD and TEM, respectively. Figure 3 shows the SEM images of the obtained film at low and high magnification. As it can be seen, the film was composed of abundant nanosheets which were interwoven with each other. The thickness of the film is about 500 nm which is equal to the width of nanosheet. The thickness of the nanosheet is about 20 nm. Figure 4 shows the XRD pattern of the ZnO nanosheets film. All the diffraction peaks match to those characteristic of the wurtzite

IOP Conf. Series: Materials Science and Engineering **761** (2020) 012008 doi:10.1088/1757-899X/761/1/012008

ZnO structure (JCPDS 036-1451). No characteristic peaks of other chemical compounds, such as Zn or Zn $(OH)_2$, were observed. The insets of figure 4 show the TEM (i) and HRTEM (ii) images of individual ZnO nanosheet. From the TEM image, it can be seen that there are a plenty of pores existed in the surface of the nanosheet. The pore size is less than 10 nm. The HRTEM analysis of an individual nanosheet reveals clear lattice fringes, with a lattice spacing of 0.26 nm corresponds well to the d-spacing of the (002) plane of the hexagonal wurtzite ZnO.

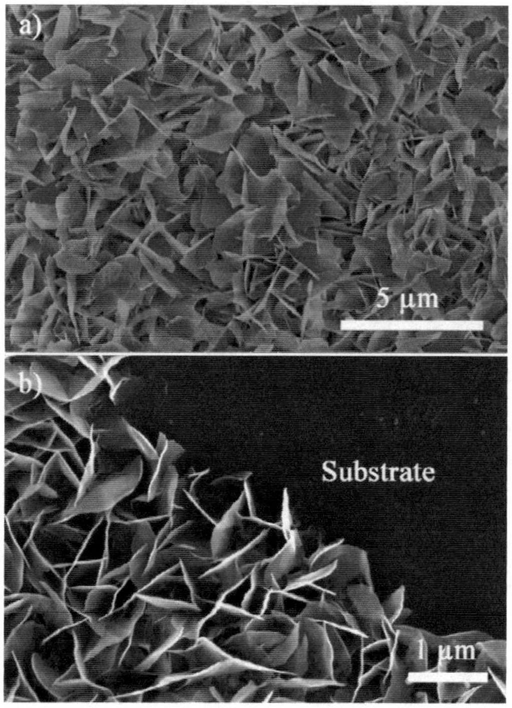

Figure 3. (a) Low- and (b) high-magnification SEM images of the ZnO nanosheets film on PET substrate.

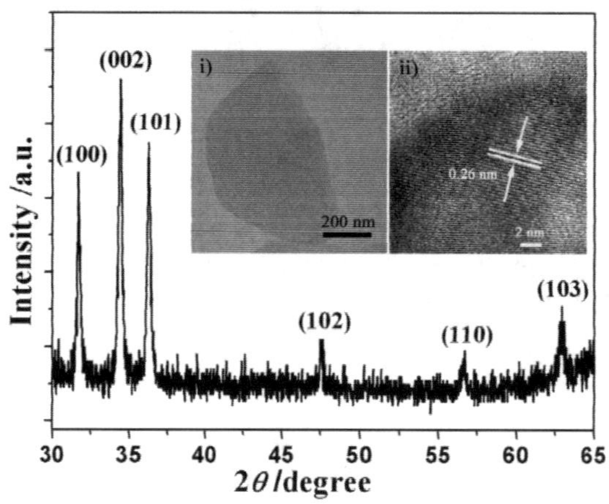

Figure 4. XRD patterns of synthesized ZnO nanosheets film. The insets show the TEM image of individual ZnO nanosheet (i) and corresponding HRTEM image (ii).

During the electrodeposition process, OH^- formed on the electrode surface according to the reaction equation as $NO_3^- + H_2O + 2e \rightarrow NO_2^- + 2OH^-$ [11], and meanwhile Zn^{2+} were constantly migrated to the cathode under an electric field. When the concentration of Zn^{2+} and OH^- reached the deposition condition of ZnO, the nanosheet began to grow. Due to the hysteresis effect of ion transport in comparison to ion consumption, the ion concentration will decrease gradually in front of the deposit which result in the nanosheet stop grows temporarily [22]. With the gradually increasing of ion concentration, the next nanosheet growth began. In this way, ZnO nanosheets film was deposited periodically by self-oscillation.

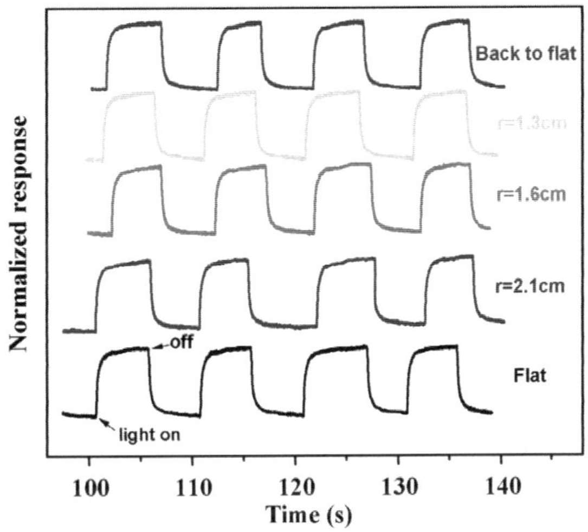

Figure 5. Time dependent photo responses of a ZnO nanosheet film device onto PET substrate subjected to different radii of curvature.

As a flexible electrical device, its electrical performance should be stable under bending conditions. In order to investigate the flexibility and stability, the device was bent to different curvature radius to assess the optoelectronic performance of the device. Figure 5 shows the photoconductor response switching behavior of a representative ZnO photoresponse device at constant bias voltages (5 V) before, during, and after it was subjected to bending at various radii of curvature (r=1.3, 1.6 and 2.1 cm) by taping the PET substrates to common glass vials (as shown in figure 2). It can be observed that the photocurrent and response/recover characteristics of the device were nearly unchanged, which reveals that the performance of the device can be hardly affected by bending. It means that there is no significant device deterioration or obvious ZnO film cracking and nanosheets peeling in these bending tests. After the above study, we can draw that the nanosheets are strong adhesion to the PET substrate, making such kind of electrodeposition based assembly approach a viable route in fabricating flexible devices.

4. Conclusions

We report a simple and effective strategy to assemble nanostructures onto plastic substrate by using ultrathin layers electrodeposition technique; here as manifested by the fabrication of ZnO nanosheets film onto PET substrate using this method. We have exploited such assembled structures for flexible electronics by the fabrication of photoresponse device which show stable photoresponse under bending at various curvature radii (r= 2.1, 1.6 and 1.3 cm). This method can also be applied in other nanostructures through electrodeposition, which can stimulate the development of low-cost, flexible and wearable technologies, limited not only to electrical devices but also to sensor applications, *etc*.

5. Conflict of interest

None.

6. References

[1] Liu Z, Xu J, Chen D and Shen G 2015 *Che. Soc. Rev.* **44** 161-192
[2] Webb R , Bonifas A , Behnaz A, Zhang Y, Yu K, Cheng H, Shi M, Bian Z, Liu Z, Kim Y, Yeo W, Park J, Song J, Li Y, Huang Y, Gorbach A and Rogers J 2013 *Nature Mater.* **12** 1078-78
[3] Hu P, Wang L, Yoon M, Zhang J, Feng W, Wang X, Wen Z, Idrobo J, Miyamoto Y, Geohegan D and Xiao K 2013 *Nano Lett.* **13** 1649-54.
[4] Bai S, Wu W, Qin Y, Cui N, Bayerl D and Wang X 2011 *Adv. Func. Mater.* **21** 4464-69
[5] Yang L, Wu L, Wu M, Xin G, Lin H and Ma T 2010 *Electroche. Commun.* **12** 1000-03
[6] Kim B, Lee S, Kang M, Ahn J and Cho J 2012 *ACS Nano.* **6** 8646-51
[7] Grimm D, Bof Bufon C, Deneke C, Atkinson P, Thurmer D, Schäffe F, Gorantla S, Bachmatiuk A and Schmidt O 2013 *Nano Lett.* **13** 213-218
[8] Yuan L, Lu X, Xiao X, Zhai T, Dai J, Zhang F, Hu B, Wang X, Gong L, Chen J, Hu C, Tong Y, Zhou J and Wang Z 2012 *ACS Nano* **6** 656-661
[9] Duan X, Niu C, Sahi V, Chen J, Parce J, Empedocles S and Goldman J 2003 *Nature* **425** 274-278
[10] McAlpine M, Friedman R, Jin S, Lin K, Wang W and Lieber C 2003 *Nano Lett.* **3** 1531-35
[11] Cui G, Gao L, Yao B, Wang S, Zhang P and Zhang M 2013 *Electroche. Commun.* **30** 42-45
[12] Xiao C, Yang T, Chuai M, Xiao B and Zhang M 2018 Phys. *Chem. Chem. Phys.* **18** 325-330
[13] Wu Z, Li H, Xiong X, Ma G, Wang M, Peng R and Ming N 2009 *Appl. Phys. Lett.* **94** 041120
[14] Wang M, Zhong S, Yin X, Zhu J, Peng R, Wang Y, Zhang K and Ming N 2001 *Phys. Rev. Lett.* **86** 3827-30
[15] Xiao C, Xiao B, Wang Y, Zhang J, Wang S, Wang P, Yang T, Zhao R, Yu H, Li Z and Zhang M 2015 *Rsc. Adv.* **5** 17945-52
[16] Cui G, Li Z, Gao L and Zhang M 2012 *Phys. Chem. Chem. Phys.* **14** 16321-25
[17] Wang K, Niu L, Zong Z, Zhang M, Wang C, Shi X, Men Y and Zou G 2008 *Cryst. Growth & Design* **8** 442-445
[18] Zhang M, Wang M, Zhang Z, Zhu J, Peng R and Ming N 2004 *Electrochimica Acta* **49** 2379-83
[19] Cui G L, Zhang M and Zou G 2013 *Scientific Reports* **3** 1250
[20] Wang Y, Cao Y, Wang M, Zhong S, Zhang M, Feng Y, Peng R, Hao X, Ming N 2004 *Phys. Rev. E* **69** 021607
[21] Liu T, Wang S, Shi Z, Ma G, Wang M, Peng R, Hao X and Ming N 2007 *Phy. Rev. E* **75** 051606
[22] Zong Z, Zhang M, Lu H, Xu D, Wang S, Tian H, Liu C, Guo H, Gao H and Zou G 2010 *Appl. Phys. Lett.* **96** 143113

IOP Conf. Series: Materials Science and Engineering **761** (2020) 012009 doi:10.1088/1757-899X/761/1/012009

Molecular Dynamics Simulation on Dielectric Constant and Thermal Conductivity of Crosslink Epoxy/ functionalized graphene Nano-composites

L Shen[1], L Zou[1], M Ding[1], T Zhao[1], L Zhang[1] and Q Li[2]

[1] School of Electrical Engineering, Shandong University, No.17923 Jingshi Road, Jinan 250061, China
[2] the State Key Laboratory of Alternate Electrical Power System with Renewable Energy Sources, North China Electric Power University, No.2 Beinong Road, Changping District, Beijing 102206, China

zouliang@sdu.edu.cn

Abstract. Nanoscale graphene fillers are expected to enhance electronic and thermal properties of polymers owing to the superior performances of graphene. In this paper, dielectric constant and thermal conductivity of crosslink epoxy/ functionalized graphene nano-composites were investigated using molecular dynamics simulations. The results show that graphene nanoplatelets (GN) can improve performances of polymer matrix, covalent functional groups can further enhance this improvement. Among different functional groups, hydroxyl is more effective than carboxyl and amino in reducing dielectric constant. Correspondingly, the introduction of graphene fillers also leads to an increase in thermal conductivity. GN and amino functionalized GN raise thermal conductivity of crosslink epoxy by 30.69% and 44.86% respectively. All the results demonstrate that functionalization of GN can suppress polarization and enhance thermal transportation.

1. Introduction

Owing to the remarkable electronic, mechanical and thermal properties of graphene, polymer based nano-composites with graphene fillers have attracted significant attention in the past few years [1-4]. A large amount of research has indicated that electronic and thermal properties of such nano-composites could be enhanced, thus they hold great potential in a wide range of practical application [5-7]. One of the most promising applications is in insulation material. Insulation material demand for low dielectric constant and high thermal conductivity. In sharp contrast to the low thermal conductivity of common insulation material such as crosslink epoxy (EP), graphene has extremely high thermal conductivity in the range of 2000-5300 $Wm^{-1}K^{-1}$ [4,8]. Dispersing graphene fillers into polymer is a promising approach to improve performances of insulation material.

When graphene is dispersed into polymer, the resultant composite performances are influenced by interfacial interaction between graphene and polymer [7]. Previous research found that covalent functional groups can enhance interaction between molecules and improve performances of nano-composites. Jen-Yu Wang et al. performed experiments and discovered that graphene oxide can not only enhance mechanical properties of polyimide but also reduce dielectric constant of polyimide by 39.4% [9]. Using Molecular Dynamics (MD), Yu Wang et al. discovered that the thermal resistance

IOP Conf. Series: Materials Science and Engineering **761** (2020) 012009 doi:10.1088/1757-899X/761/1/012009

between epoxy and graphene functionalized by carboxyl and hydroxyl has decreased by 54.8% and 31.2% respectively compared to pristine graphene [10].

Using experiments to evaluate the relationship between structure and properties of nano-composites is challenging because of the associated variables such as the extent of conversion, extent of dispersion, and experimental temperature stc. An alternative method: MD simulation was applied throughout this study. MD simulation allows for an atomic-scale description of the interactions between graphene and polymer matrix. Thereby, it has been widely used in research on nano-composites. Upon analyzing the results from simulation system, an evaluation upon nano-composites' performances can be given.

The purpose to the present article is to explore dielectric constant and thermal conductivity of EP doped with graphene nanoplatelets (GN). Models of unmodified epoxy (EP) and epoxy doped with GN (EP-GN) were firstly established. Then GN fillers functionalized by hydroxyl (-OH), carboxyl (-COOH) and amino (-NH_2) functional groups was established and dispersed into EP. This is followed by the calculation of dielectric constant and thermal conductivity of the five models respectively. Performances of nano-composites were compared to the performances of EP and amongst themselves to study the influence of different fillers to the polymer matrix. We close the paper by summarizing the results in the conclusion part.

2. Preparation of model structures

Establish model of Biphenyl A epoxy resin. The monomer and curing agent of biphenyl A epoxy resin are diglycidyl ether of bis phenol A (DGEBA) and 3,3′-diaminodiphenyl sulfone (33DDS). Molecular structures of DGEBA and 33DDS are shown in Figure 1. The model was built and structurally optimized using Material Studio software. The intra- and intermolecular forces that governed the behavior of the molecular models were described by CAMPASSII [11] force field.

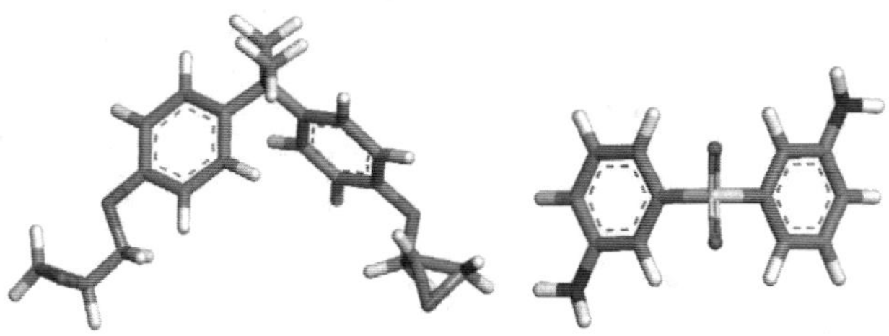

(a)DGEBA (b)33DDS

Figure 1. Molecular structures of DGEBA and 33DDS.

Models of dopants were established afterwards, including graphene nanoplatelets (GN), carboxyl functionalized graphene nanoplatelets (GNCOOH), hydroxyl functionalized graphene nanoplatelets (GNOH), and amino functionalized graphene nanoplatelets (GNNH_2). Generally, hydroxyl groups are attached to the basal plane of graphene nanoplatelets whereas carboxyl and amino groups are decorated at the edge sites [12]. The models of GN, GNCOOH, GNOH, and GNNH_2 are displayed by Figure 2.

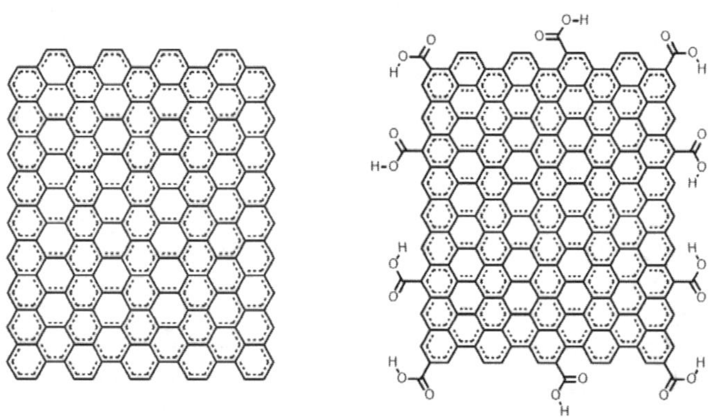

IOP Conf. Series: Materials Science and Engineering **761** (2020) 012009 doi:10.1088/1757-899X/761/1/012009

(a) model of GN (b) model of GNCOOH

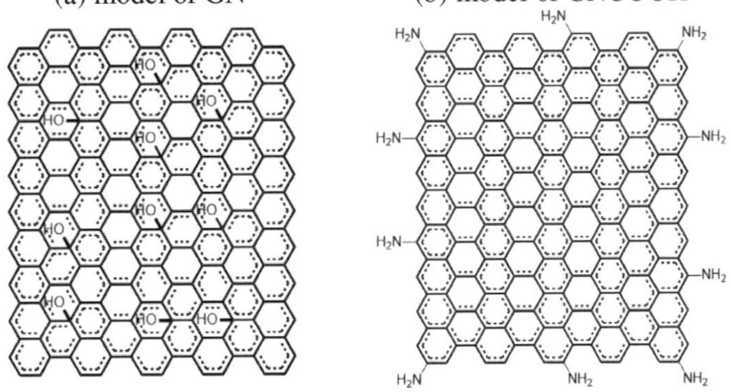

(c) model of GNOH (d) model of GNNH2

Figure 2. Models of pristine graphene and graphene covalently bonded with different functional groups.

In this simulation, 50 DGEBA and 25 33DDS were placed into a simulation box. In the cross-linking process, the scattered epoxy monomer and curing agents formed chemical bonds with each other. Next, the four kinds of dopants were placed respectively into the simulation box with the uncrosslink epoxy monomer and curing agents. Models of the four nano-composites were established repeating the cross-linking process. The following abbreviations were used to represent the nano-composite models in the paper: EP-GN, EP-GNCOOH, EP-GNOH, and EP-$GNNH_2$ systems for EP doped with GN, GNCOOH, GNOH, and $GNNH_2$. Parameters of the nano-composites are given in Table 1 and the models are displayed by Figure 3.

The established models of EP and nano-composites were relaxed in an NPT ensemble. The temperature, pressure, and time settings of the relaxation process were 298K, 1 atm, and 100ps respectively. Structures of the crosslink models were closer to the actual situation after relaxation. The resultant models were adopted as test objects in subsequent simulation. The simulation was performed in the open-source large-scale atomic-molecular massively parallel simulator (LAMMPS) [13].

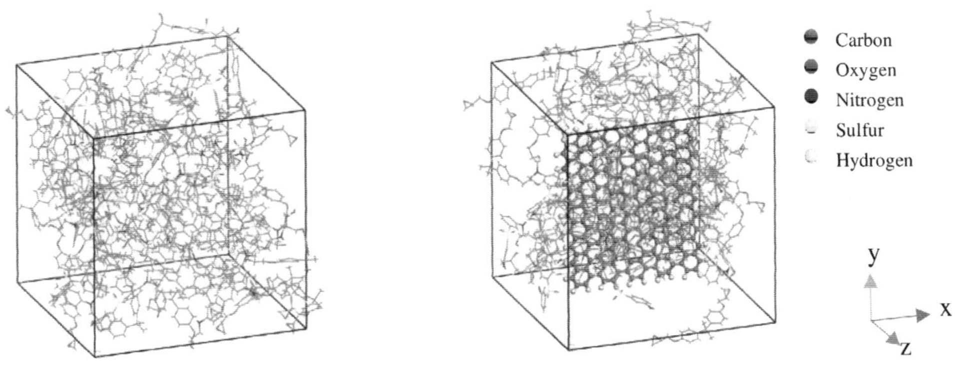

(a) Model of EP (b) Model of EP-GN

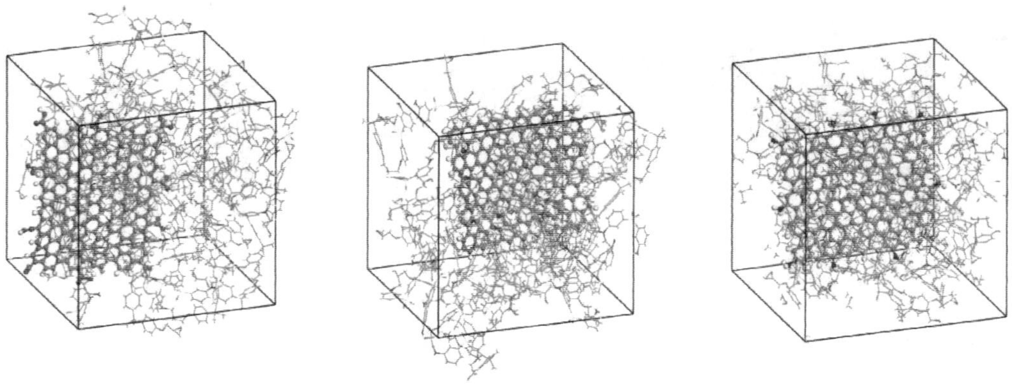

(c) Model of EP-GNCOOH (d) Model of EP-GNOH (e) Model of EP-$GNNH_2$

Figure 3. Model of Nano-composites.

Table 1. Parameters of Models.

Model	Number of Atoms in models					Molecular Weight	Mass Fraction of Dopants in Nano-composites
	C	H	N	O	S		
EP	1350	1500	50	250	25	23228.4	
EP/GN	1548	1538	50	250	25	25644.9	9%
EP/GNCOOH	1558	1539	50	270	25	26086	11%
EP/GNOH	1548	1548	50	260	25	25815	10%
EP/$GNNH_2$	1548	1549	60	250	25	25796.1	10%

3. Calculation of dielectric constant

In principle, dielectric constant can be calculated by measuring total dipole moment *M* of the model. During the simulation process, the five models were firstly relaxed in an NVT ensemble at 300K for 50ps. Then the relaxed models ran MD simulation in an NVT ensemble at 300K for 50ps. The simulation was performed for 10 times repeatedly. For each simulation, total dipole moment of model was obtained, dielectric constant ε can be described by [14]

$$\varepsilon = 1 + \frac{1}{3Vk_B T\varepsilon_0}(\langle \vec{M}^2\rangle - \langle \vec{M}\rangle^2) \tag{1}$$

V and *T* are volume and temperature of the models. k_B is the Boltzmann's constant. ε_0 is the dielectric constant of vacuum. $\langle \vec{M}^2\rangle$ is the average value of the dipole moment squared, $\langle \vec{M}\rangle^2$ is the square of the mean dipole moment values. The dielectric constants obtained by repeating the calculation of each model were averaged, the resultant dielectric constants of the five models are shown in Figure 4.

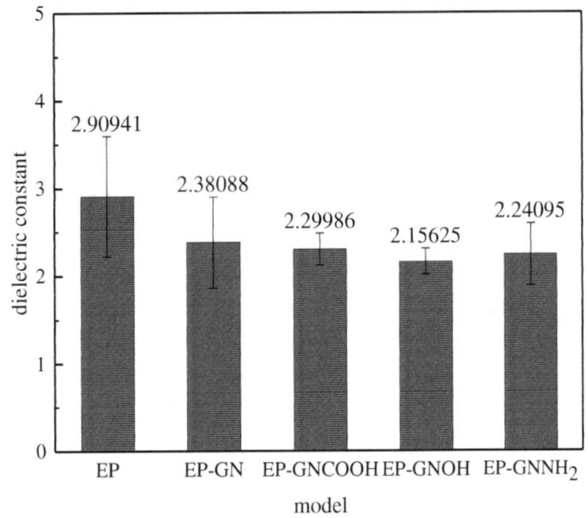

Figure 4. The dielectric constant of five nano-composite models.

The dielectric constant of EP was calculated as 2.909. This value is in good agreement with 2.5-6.0 as observed by other researchers. Dielectric constants of nano-composites generally have lower dielectric constant than EP. EP-GNOH has the lowest dielectric constant of 2.156, which is 25.9% lower than dielectric constant of EP and 9.4% lower than dielectric constant of EP-GN.

This phenomenon can be explained by the polarization mechanism of epoxy. The polarization mechanism of epoxy is dipole polarization, which depends on the mobility of chain segments. The sp^2-carbons in GN can form strong π–π stacking interactions with aromatic rings in EP. The interactions intensified the constraints on chain segments in EP-GN and reduced its dielectric constant. Molecular structure of hydroxyl can introduce strong hydrogen bonds. The bonds enhanced the constrains on chain segments and further reduced dielectric constant of EP-GNOH.

4. Calculation of thermal conductivity

A common method to calculate thermal conductivity in MD simulation is nonequilibrium molecular dynamics (NEMD) [15]. In NEMD, a heat flux is imposed on the simulation system and the thermal conductivity is related to the resultant temperature gradient by:

$$\kappa = -\frac{J_x}{\partial T / \partial x} \tag{2}$$

J_x is the heat flux which is defined as the amount of energy transferred in a given time through a surface of given area perpendicular to the flux direction (x direction in this case). $\partial T / \partial x$ is the temperature gradient along x direction.

The models were firstly heated integrally in NVT ensemble at 300K for 50ps. Then the models were heated partially in NVE ensemble for 100ps. In the partial heating process, the simulation box was divided into 20 "slices" along x direction. The slices at the two end of the box were set as fixed layer displayed by Figure 5. A "hot" plate and a "cold" plate were artificially placed next to the two fixed layer. The hot plate and cold plate were also known as heat source and heat sink. In each time step, kinetic energy of the hottest atom in the heat sink exchange with kinetic energy of the coldest atom in the heat source. The exchanging process produced a heat flux from the heat source to the heat sink, which led to a temperature gradient in the intervening region. Temperature of the slices were calculated at each iteration. Once the model reached a steady state, thermal conductivity of the material can be calculated.

IOP Conf. Series: Materials Science and Engineering **761** (2020) 012009 doi:10.1088/1757-899X/761/1/012009

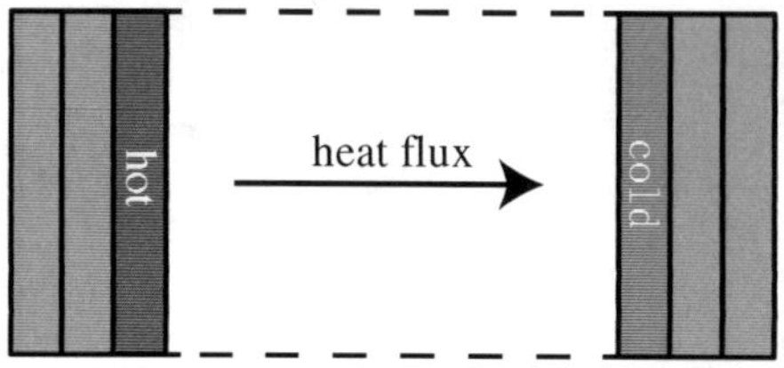

Figure. 5 Model schematic diagram of NEMD

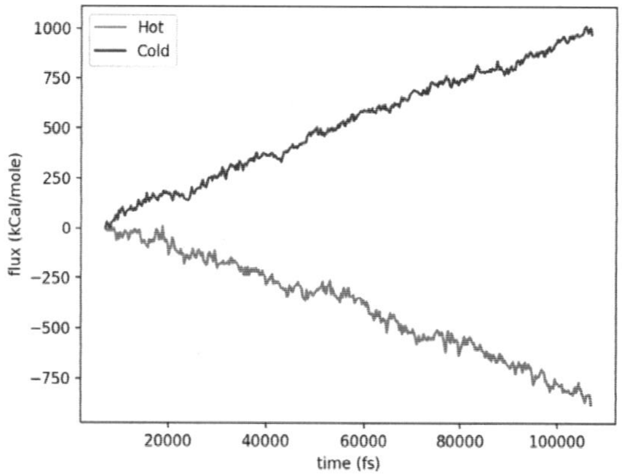

Figure 6. The energy flowing into heat source and heat sink

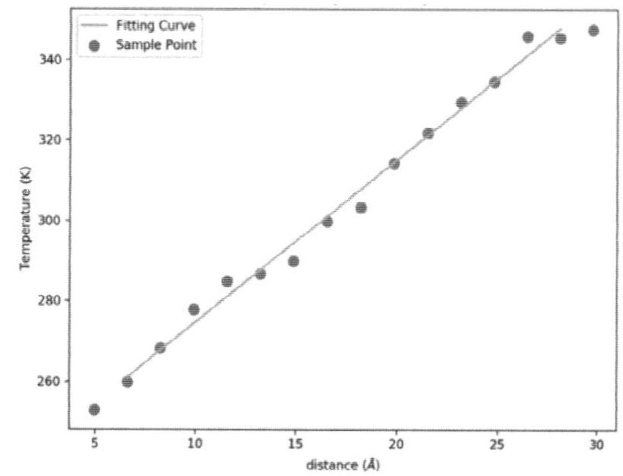

Figure 7. Temperature gradient diagram with the change of simulation time

Take x direction of EP model as an example, heat flow with the change of simulation time is displayed by Figure 6. When a stable, linear temperature profile is reached, temperature gradient of the model is displayed by Figure 7.

Repeat the above-mentioned process on the five models, for each model the simulation was performed in x, y, and z direction. Each calculation was performed for five times with a spontaneous initial energy of the molecule. Take the average of five calculation results as the thermal conductivity. The integral thermal conductivity of the model was represented by the arithmetic mean of the thermal conductivity in three directions. The results are shown in Table 2.

Table 2. The thermal conductivity of five epoxy models.

thermal conductivity κ/Wm-1K-1	x direction	y direction	z direction	integral
EP	0.1484	0.1551	0.1285	0.1440
EP-GN	0.1988	0.2647	0.1009	0.1882
EP-GNCOOH	0.2060	0.2744	0.1207	0.2004
EP-GNOH	0.2619	0.1537	0.1661	0.1939
EP-$GNNH_2$	0.2601	0.2585	0.1073	0.2086

The thermal conductivity of EP was calculated as 0.1440 $Wm^{-1}K^{-1}$. This is in reasonable disparity with the value of 0.19 $Wm^{-1}K^{-1}$ given by experiments [16]. Nano-composites with dopants have larger thermal conductivity compared to EP. GN, GNCOOH, GNOH, and $GNNH_2$ enhanced thermal conductivity of EP by 30.69%, 39.17%, 34.65% and 44.86% respectively.

It can be concluded from the results that thermal conductivity of z direction in nano-composites is smaller than thermal conductivity of x and y direction. This can be attributed to interfacial thermal resistance (also known as Kapitza resistance) cause by the large distinction of thermal properties between graphene and EP [17]. Covalent functional groups in nano-composites can enhance interaction between dopants and polymer matrix as well as diminish interfacial thermal resistance. Hence, thermal conductivity of nano-composites with functional groups was increased.

5. Conclusion

In this study, MD simulation was performed to investigate the dielectric constant and thermal conductivity of five crosslink models. The five models include crosslink epoxy and four epoxy based nano-composites. The model of crosslink epoxy was built as comparison to the nano-composites. The fillers of nano-composites were graphene nanoplatelets, carboxyl functionalized graphene nanoplatelets, hydroxyl functionalized graphene nanoplatelets, and amino functionalized graphene nanoplatelets. The influence of different dopants to the performances of nano-composites were examined.

- Calculate the dielectric constant by counting dipole moment of the simulation system. Among the different covalent groups, hydroxyl is the most effective one in reducing dielectric constant, lowering dielectric constant by 25.9% compared to EP.
- Calculate the thermal conductivity applying NEMD method. All nano-composites had different degree of enhancement in thermal conductivity. EP-$GNNH_2$ had the largest thermal conductivity which is 44.86% larger compared to thermal conductivity of EP.

The present findings provide a review on dielectric constant and thermal conductivity of epoxy doped with functionalized graphene nanoplatelets. The results can offer a useful guidance for the future development of graphene-epoxy nano-composites in insulation material.

6. References

[1] A. K. Geim, and K. S. Novoselov 2007 *Nat. Mater.* **6** 183−191
[2] C. Lee, X. Wei, J. W. Kysar and J. Hone 2008 *Sci.* **321** 385−388
[3] T. Y. Ng, J. J. Yeo, Z. S. Liu 2012 *Carbon* **50** 4887−4893
[4] A. A. Balandin, S. Ghosh, W. Bao, I. Calizo, D. Teweldebrhan 2008 *Nano Lett.* **8** 902−907.
[5] H. X. Chang, H. K. Wu 2013 *Sci.* **6** 3483−3507
[6] Y. Q. Sun, G. Q. Shi: J. Polym. Sci., Part B 2013 *Polym. Phys.* **51** 231−253
[7] J. R. Potts, D. R. Dreyer, C. W. Bielawski and R. S. Ruoff 2011 *Polymer* **52** 5−25
[8] S. Ghosh, I. Calizo, D. Teweldebrhan, E. P. Pokatilov, D. L. Nika, A. A. Balandin, W. Bao, F. Miao, C. N. Lau 2008 *Appl. Phys. Lett.* **92** 151911
[9] J. Y. Wang, S. Y. Yang, Y. L. Huang, H. W. Tien, W. K. Chin and C. C. M. Ma 2011 *J. Mater. Chem.* **21** 13569-13575.
[10] Y. Wang, C. H. Yang, Q. X. Pei and Y. Y. Zhang 2016 *Appl. Mater. Interfaces* **8** 8272−8279
[11] H. Sun, Z. Jin, C. Yang, et al. 2016 *J. Mol. Model.* **22** 47
[12] A. Lerf, H.Y. He, M. Forster, and J. Klinowski 1998 *J. Phys. Chem. B.* **102** 4477-4482.
[13] S. Plimpton 1995 *J. Comput. Phys.* **117** 1-19
[14] M. Neumann 1983 *Mol. Phys.* **50** 841-858
[15] F. Muller-Plathe 1997 *J. Chem. Phys.* **106** 6082−6085
[16] K. Chu, W. S. Li, H. Dong 2013 *Appl. Phys. A* **111** 221−225
[17] E. Swartz, R. Pohl 1989 *Rev. Mod. Phys.* **61** 605– 668

Acknowledgments

This study is supported by the State Key Laboratory of Alternate Electrical Power System with Renewable Energy Sources(Grant No.LAPS19006).

Preparation of Nano-Graphite Oxide

LUO Xue-ping[1, 2] and lAI Qi[1, 3]
1. Lab of Graphene, Panzhihua University, Panzhihua 617000, China
2. College of College of foreign languages, Panzhihua University, Panzhihua 617000
3. College of titanium and vanadium, Panzhihua University, Panzhihua 617000

Abstract. In order to get high yield of nano-Graphite oxide through raw materials of nature graphite, the preparation process of oxidation, expansion, and ultrasound smash were studied. It is found that the more intense of oxidation reaction is, the more crack the expanded graphite is, and more completely the expansion of the nano-Graphite oxide will be. The Large expansion volume of the nano-Graphite oxide can be more easily crushed into nano-Graphite oxide at the same ultrasonic time. And the longer the ultrasonic time is, the more the yield of nano particle size of expanded nano-Graphite oxide is. And the less ultrasonic time can get the smaller nano particle size. The result shows that the best process conditions of nano-Graphite oxide are as follows: $V_{sulfate\ acid}$: $V_{Nitrate\ acid}$=5:1, m (fine graphite): m (Potassium permanganate) =1:0.18. The expansion temperature is 950-1000°C, and ultrasonic time is 10h.

1. Introduction
The development of various methods for producing nano-Graphite oxide has stimulated a vast amount of research in recent years[1]. The remarkable properties of nano-Graphite oxide (including graphene) reported so far include high values of its excellentperformance[2]. Graphite are promising candidates as components in applications such as energy-storage materials[3], paper-like materials[4], polymer composites. Since it had prepared, graphite oxide has been mainly produced by the Brodie, Staudenmaier and Hummers methods[5]. All three methods involve oxidation of graphite in the presence of strong acids and oxidants. The level of the oxidation can be varied on the basis of the method, the reaction conditions and the precursor graphite used. Although chemists had enjoyed most success with the production of chemically modified nano-Graphite oxide, the yield of nano-Graphite oxide is a big problem. So the intercalation and expansion of the preparation of graphene were studied in order to get higher yield of nano-Graphite oxide in this research.

2. Experiment
A quantity of about 163μm nature flake graphite, concentrated different condition of acid and potassium permanganate were stirred in the flask. After 60 min, the GIC was prepared. It was then washed in water to neutrality, dehydrated and dried at a temperature of 60 °C. Then the GIC was rapidly heated in a tubular furnace kept at 1000 °C for 10 s for expanding.

Nature graphite in this study is divided into two groups according to its particle sizes: (a) 420μm and (b) 420-250μm is large scale graphite (taken from BEISHU Graphite Co. Shandong Province, China), (c) 250-149μm and (d) 149-85μm is fine flake graphite (taken from PANXI Graphite Co. Sichuan Province, China). The designations (a, b, c,d) are used throughout the manuscript. Other reagents include sulfuric acid (AR), nitrate acid (AR), and potassium permanganate (AR). The graphite of (a), 9b), (c) and (d) are all received the same processing, including pretreatment, oxidization and expansion at temperature of 1000°C.

IOP Conf. Series: Materials Science and Engineering **761** (2020) 012010 doi:10.1088/1757-899X/761/1/012010

Expanded graphite was put in to the solution of alcohol (70%) and sonicated into nano particle size. The ultrasonic time is 4h, 6h, 8h and 10h. And then the nano particle size of expanded graphite was measured by laser particle size analyzer.

3. Experimental results and analysis

3.1. The effect of intercalation specie

The data in Table 1 shows that when oxidation is strong, the expansion volume is so large [3]. When the amount of potassium permanganate is too large, the consistency of oxidants is cut down, so large expandable volume GIC could not be acquired. When the oxidation is too small, the layers of flake graphite could not be inserted by acid and Potassium permanganate completely, so The best process conditions are as follows: $V_{\text{sulfate acid}}$: $V_{\text{Nitrate acid}}$=5:1, m (fine graphite): m (Potassium permanganate) =1:0.18.

Table 1. The expansion volume of GIC made from different specie

Intercalation agent	Sulfate acid: Nitrate acid =5:1 The ratio of Potassium permanganate is as follows				
	0	6	12	18	24
expansion volume(ml/g)	92	164	336	411	318

Figure 1 shows the XRD patterns of GICs. With the oxidation reaction conditions increased, a new diffraction peak from 25.35° to 26.21° emergence, and the intensity of diffraction also on the rise, indicating the high oxidation attributes to the high intercalation, corresponding the d-space from 0.3374nm to 0.3480nm.

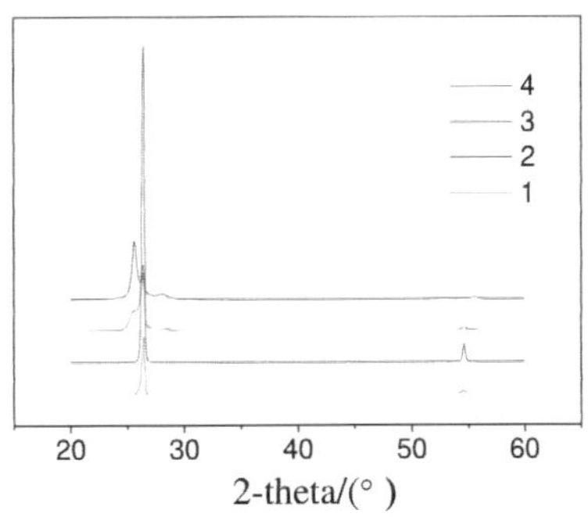

Figure 1. XRD of the GIC with different intercalation degree

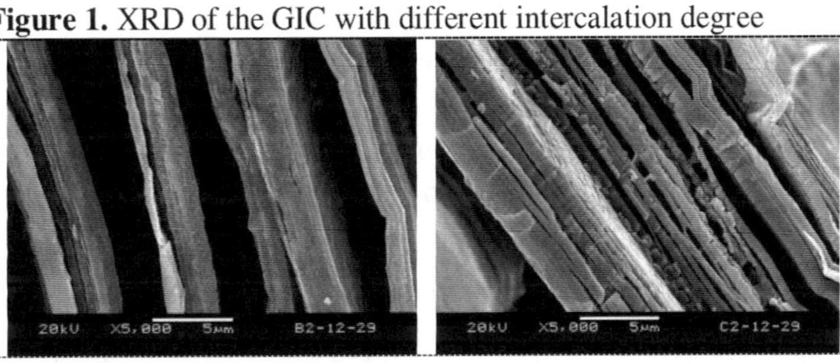

Figure 2. SEM of GIC under different oxidation reaction conditions

The SEM of the GIC (Figure 2) shows the difference in graphite oxide prepared under different oxidation reaction conditions. The more intense the oxidation reaction is, the more the crack graphite is, and more completely the expansion of the graphite oxide will be.

3.2. Effect of temperature

Figure 3 shows the expansion volume of four kinds of graphite (a, b, c and d) at different temperatures in the same condition. It is found that the higher the temperature is, the higher the expansion volume of GIC is. When the temperature is reached to 950°C, the expansion volume is higher. But if temperature and other conditions are the same, the larger flake graphite will be of higher expansion volume. Namely, the expansion volume of two kinds of large scale graphite (a and b) is higher than that of two kinds of fine flake graphite(c and d) at the same temperature. According to this fact it is concluded that in order to achieve satisfied expansion volume, fine flake graphites needs higher temperature compared with large flake graphite. When the temperature is 1150°C, the expansion volume of fine flake graphite with 250-149μm can be reached 341ml/g and the graphite with 149-85μm can be reached 230ml/g. the Large expansion volume of the graphite can be more easily crushed into nano-Graphite oxide at the same ultrasonic time. So the expansion temperature is 950-1000°C.

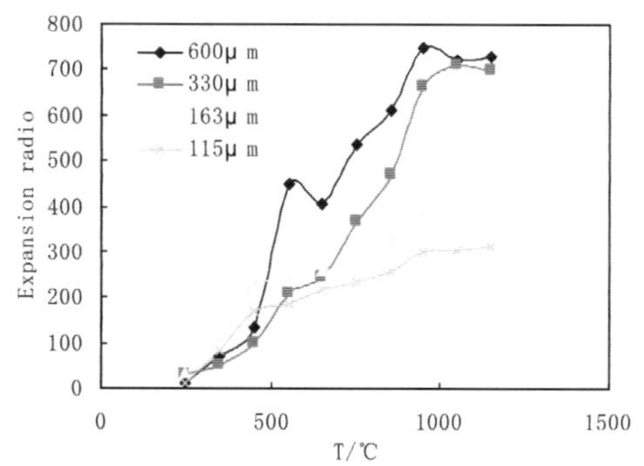

Figure 3. The expansion volume of nano-Graphite oxide with different temperature

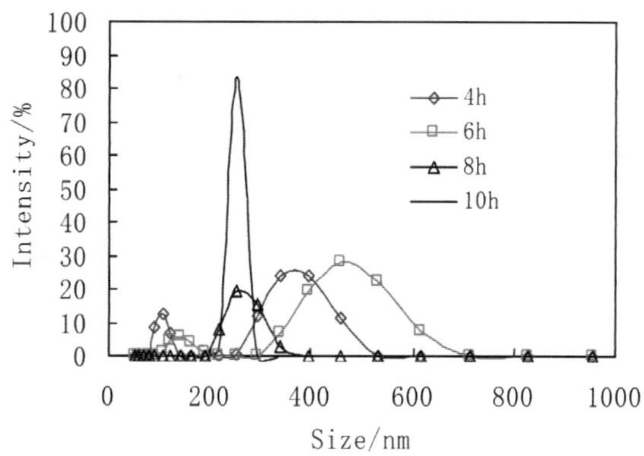

Figure 4. The size of Expanded nano-Graphite oxide with different ultrasonic time

3.3. Effect of ultrasonic time

Figure 4 shows the effect of ultrasonic time. Expanded nano-Graphite oxide was sonicated into nano particle size. The longer the ultrasonic time is, the more the yield of nano particle size of expanded graphite is. When ultrasonic time is 10h, the particle size is range from 200nm to 300nm. But the less ultrasonic time can get the smaller nano particle size. When ultrasonic time is 4h, the smaller particle size can reach to 50nm.

4. Conclusion

Nano-Graphite oxide can be prepared by oxidation, expansion, and ultrasound smash. Oxidation is more stronge, the expansion volume is more large. The Large expansion volume of the graphite can be more easily crushed into nano-Graphite oxide at the same ultrasonic time. The longer the ultrasonic time is, the more the yield of nano particle size of expanded graphite is. And the less ultrasonic time can get the smaller nano particle size. The best process conditions of nano-Graphite oxide are as follows: $V_{\text{sulfate acid}}$: $V_{\text{Nitrate acid}}$=5:1, m (fine graphite): m (Potassium permanganate) =1:0.18. the expansion temperature is 950-1000°C, and ultrasonic time is 10h.

5. References

[1] Geim A K, Novoselov K S. The rise of graphene. Nature Mater [J], 2007, 6(3): 183–191.

[2] Lee C., Wei X., Kysar J W, et al. Measurement of the elastic properties and intrinsic strength of monolayer graphene. Science [J], 2008, 321: 385-388
[3] Stoller M D, Park S, Zhu Y, et al.. Graphene-based ultracapacitors. Nano Lett [J], 2008, 8: 3498-3502
[4] Dikin Dmitriy A. Stankovich Sasha, Zimney Eric J, et al. Preparation and characterization of graphene oxide paper. Nature [J], 2007, 448(7152): 457-460.
[5] Stankovich Sasha, Dikin Dmitriy A, Dommett Geoffrey H B, et al. Graphene-based composite materials. Nature [J], 2006, 442(7100): 282-286.

IOP Conf. Series: Materials Science and Engineering **761** (2020) 012011 doi:10.1088/1757-899X/761/1/012011

Tyre Side-Slip Deformation of Dual-Steering Multi-Axle Vehicles Based on Steady-State Cornering

Yunchao Wang, Zhichao Hu, Bo Li, Zixu Li
College of Mechanical and Energy Engineering, Jimei University, Xiamen, Fujian, 361021, China

Abstract. It is well-known that the tyre side-slip deformation is a static indeterminacy problem for the full-vehicle tyres. For the dual-steering multi-axle vehicles on steady-state cornering, it is a closed and dissipative system in a stable dynamic balance state. Therefore, the minimum energy dissipation rate principle is adopted to solve the above indeterminacy problem. Moreover, a new method is proposed to develop a formula for calculating the tyre side-slip angles of the vehicles with dual steering systems (DSS). The validity of the formula is examined using the formula from literature and simulation results of the reference model

1. Introduction

In practise, multi-axle vehicles with a single steering system have some steering reliability problems. Concern for achieving better steering reliability performance has led to the use of a dual system with a hydraulic-assisting steering link mechanism system (HASLMS) for the front axles and an electrohydraulic active-steering system (EHASS) for the rear axles on these vehicles. Multi-axle vehicles such as the all-terrain cranes QAY1000 from XCMG and LTM 1400.7.1 from Liebher have become available. Figure 1 shows the chassis of the all-terrain crane LTM 1400.7.1.

Figure 1. Chassis of all-terrain crane LTM 1400.7.1.

The motion of a vehicle is governed by the forces generated between the tyre and the road. Therefore, tyre lateral force measurements are of interest to vehicle designers for stability control and ABS braking applications [1]. They are also essential to study vehicle dynamics, which have been examined by several researchers [2–4]. The lateral tyre force is generated as a function of tyre side-slip angle. However, the lateral force obtained is ultimately limited by the friction limit of the road. Because knowledge of the side-slip angle is important for vehicle stability control, real-time information about the side-slip angle is critical to the performance of integrated vehicle dynamics control (IVDC) strategies [4]. Owing to the difficulty of directly measuring the side-slip angle [6–7], its estimation accuracy becomes very important. Researchers have demonstrated that a combination of GPS and inertial sensors can provide accurate measurements of the slip angle [8–9]. However, until now, no attempt has been made with regard to side-slip-angle-estimation-based lateral dynamics control of multi-axle steering vehicles with double steering systems. This motivates the current research

2. Steady-state cornering of multi-axle vehicles with DSS

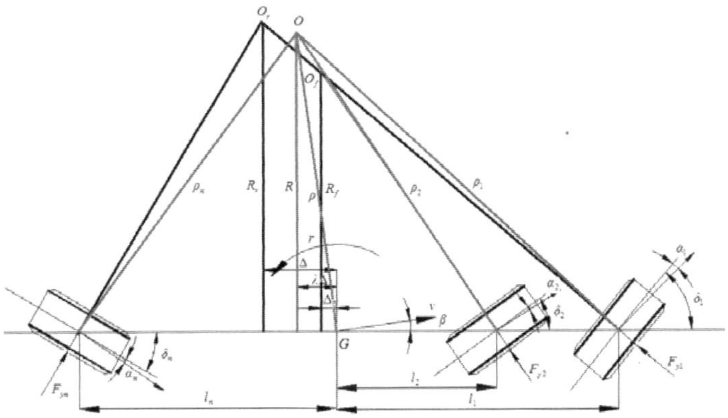

Figure 2. Dynamic model of multi-axle vehicles with DSS and coordinate systems.

Figure 2 shows coordinate systems used to describe a multi-axle vehicle with side-slip angle β, vehicle velocity v, and yaw rate r at the vehicle centre of gravity. β and r are positive in the counterclockwise direction, and v is positive in the x direction. Assuming that all of the left and right wheels are concentrated to the centre line of the vehicle, the model of the multi-axle vehicle becomes a simplified multi-axle bicycle one. The vehicle has two steering systems with HASLMS for axles 1~k and EHASS for axles (k+1)~n. The vehicle has interconnected hydropneumatic suspensions for supporting the vehicle body. Because the tyre side-slip angle and yaw rate at the centre of gravity are constant in steady-state cornering, dv/dt, $d\beta/dt$, and dr/dt become zero. Moreover, the lateral force of the side wind is neglected.

Point G is the vehicle centre of gravity, and a moving coordinate system (x, y) is attached to the point. Point O_f is the turning centre of the front wheels steered by HASLMS. Δ_0 denotes the distance between the foot point of the point O_f on the x-axis and the vehicle centre of gravity G. The numerical value of Δ_0 is positive as point O_f is in front of point G; Otherwise, it is negative. Point O_r denotes the turning centre of the rear wheels steered by EHASS. Δ denotes the distance between the foot point of point O_r on the x-axis and the vehicle centre of gravity G. The numerical value of Δ is positive as point O_r is in front of point G along the vehicle's travelling direction; Otherwise, it is negative.

The numerical value of Δ is calculated according to the active steering control algorithm of the rear wheels. l_i denotes the distance from the vehicle centre of gravity G to the ith axle in the x-direction. l_i is positive as the i_{th} axle is in front of the vehicle centre of gravity G. In fact, the actual turning centre of all wheels is the point O considering the tyre side-slip angles produced by the centrifugal force. λ is the coordination-deformation coefficient of the tyres. The distances from points O_f, O, and O_r to the x-axis are R_f, R, and R_r, respectively. F_{xi} and F_{yi} denote the x-component and y-component of the frictional forces F_i acting on the i_{th} wheel, respectively. The steering angle of each tyre is described by Δ_i. The subscript i, ranging from 1 to n, indicates the position of each tyre, starting from the front to the rear. The side-slip angle of the i_{th} tyre is α_i.

Assuming the wheel steering angle is small, for steering angles of wheels steered by SLMSHA, the ratio Hi of Δ_i to Δ_1 can be expressed as

$$\frac{l_i-\Delta_0}{l_1-\Delta_0}=\frac{\tan\delta_i}{\tan\delta_1}\Rightarrow H_i=\frac{\delta_i}{\delta_1}\approx\frac{l_i-\Delta_0}{l_1-\Delta_0} \tag{1}$$

Because the distance between points Of and Or increases as the vehicle speed increases, the steering angles of wheels steered by EHASS should be calculated according to the turning centre Or. Therefore, for the steering angles of wheels steered by EHASS, the ratio K_i of Δ_i to Δ_1 is different than the previous ratio H_i. This can be expressed as

$$\frac{l_i-\Delta}{l_1-\Delta}=\frac{\tan\delta_i}{\tan\delta_1}\Rightarrow K_i=\frac{\delta_i}{\delta_1}\approx\frac{l_i-\Delta}{l_1-\Delta} \tag{2}$$

3. Deformation of tyres with side slips of vehicles with DSS

It's well known that the side-slip angles are highly significant in maintaining the stability of the vehicle motion. The deformation distribution of tyres with side slip of an entire vehicle may provide an important way to evaluate the motion stability of the vehicle. For multi-axle vehicles, however, the calculation of side-slip angles is an indeterminate problem. Here, a new method for calculating tyre side-slip angles is introduced in detail.

According to the definition of the minimum energy dissipation rate principle, for a closed and dissipative system in a stable dynamic balance state [10], the energy consumption rate is minimum corresponding to the constraints imposed on the system. The static indeterminate problem of the deformation of tyres with side slip is to be solved by inducing this principle. Therefore, the deformation energy consumption rate of tyres with side slip caused by a cornering motion can be written as

$$P=\sum_{i=1}^{n}F_{yi}\Delta v_i \tag{3}$$

where the side-slip velocity of a tyre can be written as

$$\Delta v_i=\frac{v}{\rho}\rho_i\sin\alpha_i \tag{4}$$

where ρ_i is the turning radius of the i_{th} wheel, and ρ is that of the vehicle centre of gravity.

In the region where the side-slip angle α_i is small, $\sin\alpha_i \approx \alpha_i$, and the terms with an order of more than three of Taylor's series of α_i can be ignored. Therefore, the energy dissipation rate of the deformation of tyres with side slip can be written as

$$P=\frac{v}{\rho}\sum_{i=1}^{n}C_i\alpha_i^2\rho_i \tag{5}$$

Equation (5) is an objective function. For the deformation of tyres with side slip to correspond to real conditions, the following constraints should be satisfied:

(1)The sum of all forces in the y direction should be zero, neglecting the longitudinal forces F_x.

(2)The sum of the moments of all lateral forces around the vehicle centre of gravity G should be zero.

The constraint equations can be described as

$$\varphi_1=\sum_{i=1}^{n}C_i\alpha_i\cos\delta_i-\frac{mv^2}{\rho}\cos\beta=0 \tag{6}$$

$$\varphi_2 = \sum_{i=1}^{n} C_i l_i \alpha_i \cos\delta_i = 0 \tag{7}$$

Adopting the method of Lagrange multipliers to solve the optimization problem, the optimization function can be written as

$$L = \frac{v}{\rho}\sum_{i=1}^{n} C_i \rho_i \alpha_i^{\,2} + \eta_1\left(\sum_{i=1}^{n} C_i \alpha_i \cos\delta_i - \frac{mv^2}{\rho}\cos\beta\right) + \eta_2 \sum_{i=1}^{n} C_i l_i \alpha_i \cos\delta_i \tag{8}$$

Take a partial derivation of the function L with respect to α_i, η_1, and η_2, and make it equal to zero:

$$\frac{\partial L}{\partial \alpha_i} = 2\frac{v}{\rho} C_i \rho_i \alpha_i + \eta_1 C_i \cos\delta_i + \eta_2 C_i l_i \cos\delta_i = 0 \tag{9}$$

$$\frac{\partial L}{\partial \eta_1} = \sum_{i=1}^{n} C_i \alpha_i \cos\delta_i - \frac{mv^2}{\rho}\cos\beta = 0 \tag{10}$$

$$\frac{\partial L}{\partial \eta_2} = \sum_{i=1}^{n} C_i l_i \alpha_i \cos\delta_i = 0 \tag{11}$$

From Equation (9), the side-slip angle of the i_{th} tyre can be obtained:

$$\alpha_i = -\frac{\rho}{2v}\frac{\eta_1 \cos\delta_i + \eta_2 l_i \cos\delta_i}{\rho_i} \tag{12}$$

Substituting Equation (12) into (11):

$$\eta_1 = -\eta_2 \frac{\sum_{i=1}^{n} C_i \frac{l_i^{\,2} \cos\delta_i^{\,2}}{\rho_i}}{\sum_{i=1}^{n} C_i l_i \frac{\cos\delta_i^{\,2}}{\rho_i}} \tag{13}$$

Substituting Equation (12) and (13) into (10), and the following formula can be obtained:

$$\eta_2 = \frac{2mv^3}{\rho^2}\frac{\cos\beta \sum_{i=1}^{n} C_i l_i \frac{\cos\delta_i^{\,2}}{\rho_i}}{\sum_{i=1}^{n}\sum_{j=1}^{n} \frac{C_i C_j \cos\delta_i^{\,2} \cos\delta_j^{\,2}}{\rho_i \rho_j} l_i\left(l_i - l_j\right)} \tag{14}$$

Substituting Equation (14) into (13):

$$\eta_1 = -\frac{2mv^3}{\rho^2}\frac{\cos\beta \sum_{i=1}^{n} C_i l_i^{\,2} \frac{\cos\delta_i^{\,2}}{\rho_i}}{\sum_{i=1}^{n}\sum_{j=1}^{n} \frac{C_i C_j \cos\delta_i^{\,2} \cos\delta_j^{\,2}}{\rho_i \rho_j} l_i\left(l_i - l_j\right)} \tag{15}$$

Substituting Equations (14) and (15) into (12):

$$\alpha_i = \frac{mv^2 \cos\beta \cos\delta_i}{\rho\rho_i} \frac{\sum_{j=1}^{n} C_j l_j (l_j - l_i) \frac{\cos\delta_j^2}{\rho_j}}{\sum_{k=1}^{n}\sum_{j=1}^{n} \frac{C_k C_j \cos\delta_k^2 \cos\delta_j^2}{\rho_k \rho_j} l_k (l_k - l_j)} \tag{16}$$

4. Validation of the method

4.1. Comparison with the model from literature

To validate the method, a sample of five-axle steering vehicles with HASLMS was used for a simulation analysis, and comparisons were conducted between the calculation results from the derived formula(16)and from the formula calculating the tyre side-slip angle in [11].The formula for the tyre side-slip angle in [11]can be described as

$$\alpha_i = \frac{\sum_{i=1}^{n} C_i l_i^2 - l_i \sum_{i=1}^{n} C_i l_i}{\sum_{i=1}^{n} C_i \sum_{i=1}^{n} C_i l_i^2 - \left(\sum_{i=1}^{n} C_i l_i\right)^2} \frac{mv^2}{2\rho} \tag{17}$$

If $\rho_i \approx \rho$, $\cos\beta \approx 1$, and $\cos\Delta_i \approx 1$, Equation (16) is transformed into Equation (17). Therefore, the calculation accuracy of Equation (16)is higher than that of Equation (17), and Equation (16)is a more generalised form for tyre side-slip angles. For example, if n=2 in formula (17), it becomes the calculation formula for tyre side-slip angles β_f and β_r of two-axle vehicles, noting that l_1-l_2, C_1, C_2, l_1, and l_2 here are the equivalents of l, $2K_f$, $2K_r$, l_f, and l_r on page 66 in [12], respectively. This further validates that the calculation formula for the tyre side-slip angle derived from the proposed method is correct in theory.

4.2. Comparison with the reference model

*4.2.1 Setting of simulation parameters.*Comparative analyses of the tyre side-slip angles calculated with the two methods were performed using MATLAB software. The setting of the simulation parameters was as follows: l_1=4.07m, l_2=1.32m, l_3=-0.3m, l_4=-1.92m, l_5=-3.54, vehicle weight m=58000kg, turning radius of vehicle gravity centre calculated by Equation (9), turning radius of every axle calculated by Equation (11), cornering stiffness C=612 kN/rad for all tyres, and $\Delta = \Delta_0$=0 for the vehicle with HASLMS.

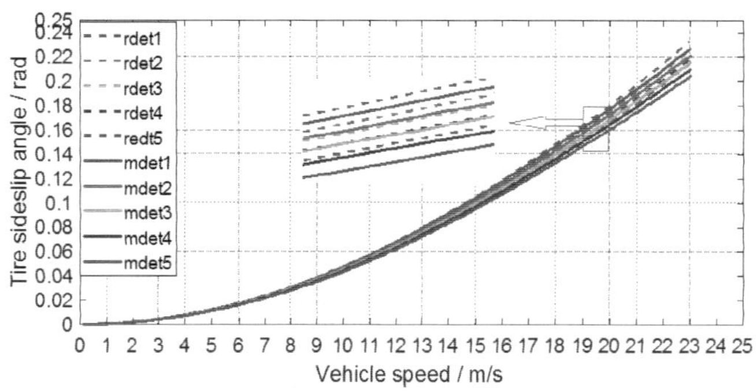

Figure 3. Comparison of results of tyre side-slip angles.

4.2.2 Analytic demonstration of simulation results by comparison. In the computer simulation, a steady-state cornering scenario was considered. The vehicle speed accelerated from 0 m/s to 23 m/s at

IOP Conf. Series: Materials Science and Engineering **761** (2020) 012011 doi:10.1088/1757-899X/761/1/012011

a constant steering angle of 0.1 rad for the first wheel. The comparison results are shown in Figure 3. The results calculated using Equation (16) and those from Equation (17) are marked with mdet* and rdet*, respectively, with the asterisk * representing the serial number of axles. The maximum discrepancy between them was less than 0.01 rad because Equation (17) is just the simplified form of Equation (16) as previously mentioned. Hence, it can be thought that they are approximately equal, thereby proving the validity of the method.

5. Conclusions

The paper presents the deformation distribution of tyres with side slip for multi-axle vehicles with DSS. A new method was proposed based on the deformation energy optimization function related to all constrained tyres. A minimum energy dissipation rate principle and Lagrange multiplier method were introduced for the method. A formula for calculating the tyre side-slip angle was derived, and it has higher calculation accuracy than that of the model from literature.

6. References

[1] Ehsan Hashemi, Mohammad Pirani, Amir Khajepour, Alireza Kasaiezadeh, Shih-Ken Chen, Bakhtiar Litkouhi, Corner-based estimation of tyre forces and vehicle velocities robust to road conditions, *Control Engineering Practice*, 2017; 61: 28–40.

[2] T. Lam, H. Qian, Y. Xu. Omnidirectional steering interface and control for a four-wheel independent steering vehicle. *IEEE/ASME Trans Mechatron*, 2010; 15: 329–338.

[3] B.Y. Li, W.H. Li, O. Kennedy, H.P. Du. The dynamics analysis of an omni-directional vehicle. *Int J Automot Technol*, 2014; 15 (3): 387–398.

[4] B. Li, H. Du, W. Li. A novel method for side slip angle estimation of omni-directional vehicles. *SAE Int J Passeng Cars: Electron Electr Syst*, 2014; 7 (2), 10.4271/2014-01-0303.

[5] Side slip angle prediction model of an off-road tyre on different terrains. *Journal of Terramechanics*, 2014; 56 (12): 25–35.

[6] P.J.T. Venhovens, K. Naab. Vehicle dynamics estimation using Kalman filters. *Vehicle Syst Dyn*, 1999; 32: 171–184.

[7] J. Stephant, A. Charara, D. Meizel. Virtual sensor: Application to vehicle sideslip angle and transversal forces. *IEEE Trans Ind Electron*, 2004; 51 (2):278–289.

[8] Robert Daily, David M. Bevly. The use of GPS for vehicle stability control systems. *IEEE Transactions on Industrial Electronics*, 2004, Vol. 51, No.2:270–277.

[9] J. Ryu, E. Rossetter, and J. C. Gerdes, Vehicle sideslip and roll parameter estimation using GPS, in Proc. *AVEC*, 2002, pp. 373–380.

[10] Xu Guobin, Lian Jijian. Theories of the minimum rate of energy dissipation and the minimum entropy production of flow (I). *J.Hydraul.Eng*. 2003; (5): 35–40.

[11] Qin Gang. Research on complex mechanical system modeling and its application in multi-axis chassis development [dissertation]. *Hangzhou: Huazhong University of Science &Technology at Hangzhou*; 2011.

[12]Masato Abe, Warren Manning. Vehicle handling dynamics (theory and application). *Elsevier Academic Press*; 2009. Chapter 2, *Tyre Mechanics*; p. 6 and Chapter 3, *Fundamentals of Vehicle Dynamics*; p. 62.

Acknowledgements

This work was supported by the National Natural Science Foundation of China under Grant [number 51575233].

IOP Conf. Series: Materials Science and Engineering **761** (2020) 012012 doi:10.1088/1757-899X/761/1/012012

Preparation and Characterization of Anatine TiO_2/Expanded Graphite with High Expansion Volume

Lai Qi[1,2], Peng Fuchang[1,2], Zhong Chanyu i[1,2] Luo Xueping[2,3] Liu Lei[2] and Xiao Chuanhai
1. Lab of Graphene, Panzhihua University, Panzhihua 617000, China
2. College of titanium and vanadium, Panzhihua University, Panzhihua 617000
3. College of College of foreign languages, Panzhihua University, Panzhihua 617000

Abstract. Element of Ti was introduced to expandable graphite surface by Sol-gel method; TiO2/expanded graphite was prepared after expansion of high temperatures. Detection by scanning electron microscope(SEM), X-ray diffraction(XRD), energy disperse spectroscopy(EDS) and differential scanning calorimetry(DSC) to semples, the results were founded: The better process conditions with high expansion volume are as follows: graphite that particle size is more than 180μm, expansion temperature is about 650°C-750°C. The contents of Ti in compounds of anatine TiO2/expanded graphite can be reached 3.16%; the volume of compounds can be reached 234ml/g. The results showed that the lower temperature and smaller particle size of graphite is benefit for anatine type of TiO2, but higher temperature is good to high expansion volume. The preparation condition should be balanced with crystal type, heat loss and expansion volume.

1. Introduction

TiO_2 photocatalyst is drug-free, stable, high catalytic activity, anti-chemical corrosion and corrosion-ray [1-2]. In many papers, carrier is discussed as a major problem, such as activated cabon [3-7], carbon fiber [8-9], et.al. Mounting of TiO_2 onto activated carbons and carbon fibers were reported to result in certain reduction of specific surface area of carbons. This was reasonably supposed to be due to the preferential deposition of TiO_2 particles at the entrance of the pores on the surface of substrate carbon. So, the expanded graphite was chosen not only because expanded graphite is flexible, compressible, and elastic, as well as being lubricious and resistant to heat and corrosion, it has been applied in large quantities in stuffing, gaskets, heat insulators, etc. [10], but also because of its oil absorption, 1g of a lump of expanded graphite sorbing more than 80g of heavy oil [11].

In the present work, therefore, TiO_2 particles were precipitated on the surface of the graphite intercalated compounds (GIC) to obtain gelation-GIC (gel-GIC), and then being followed by expansion at high temperature in order to obtain rutile TiO_2/expanded graphite with high expansion volume. And the structures of TiO_2/ expanded graphite were characterized.

2. Formatting the title, authors and affiliations

2.1. Preparation

Nature graphite in this study is divided into three groups according to its particle sizes: (a)>450μm, (b) 250-180μm, and (c) 106−85μm. Other reagents include sulfuric acid (AR), nitrate acid (AR), and potassium permanganate (AR).

A quantity of nature flake graphite with different particle size are dried at 80°C in a vacuum oven for 24h, concentrated different condition of acid and potassium permanganate are stirred in the flask.

After 60min, the mixture is carefully washed and filtrated with deionized water until the pH level of the solution reaches 6. After being dried at 60°C in a vacuum oven for 24h, the graphite intercalated compound (GIC) is prepared.

TiO_2 particles were adhesived onto the surface of GIC through the tetrabatyl orthotitanate (TBOT), triethanolamine (TEA) and ethanol, m (tetrabutyl titanate):m (triethanolamine) =1:0.4. GIC of 10 g was suspended in the solution in order to get gel-GIC. And the gel-GIC was prepared after the sol was dried at 373 K for 24 h and grounded to powder. Then the substrate GIC was expanded at high temperature for 10-15s in a muffle furnace to form expanded graphite.

2.2. Characterization

Scanning electron microscope (SEM) and energy disperse spectroscopy (EDS) studies were carried out on a VEGA's electron microscope (VEGA II XMH). X-ray Diffraction patterns were obtained by a Broker's diffractometer (D8 ADVANCE). Differential scanning calorimetry (DSC) was made by a TA's Analyzer (TA-Q20-DSC). The expansion ratio of GIC is denoted using expansion volume (ml/g).

3. Experimental results and analysis

3.1. Preparation condition

The data in Table 1 shows the expansion volume for different particle size of graphite. It is found that the expansion volume of composites is higher when large particle size is used. Although the expansion volume of gel-GIC is smaller than GIC with the same particle size, the particle size of more than 180μm of gel-GIC can be reach 200 mL/g.

Table 1. Expansion volume for different size of the gel-GICs

Size(μm)	450-850	250-180	106-85
1.143	0.285	0.286	0.593
.143	0.285	0.067	0.796
.571	0.109	0.600	0.439

3.2. Structure and microstructure analysis of gel-GIC

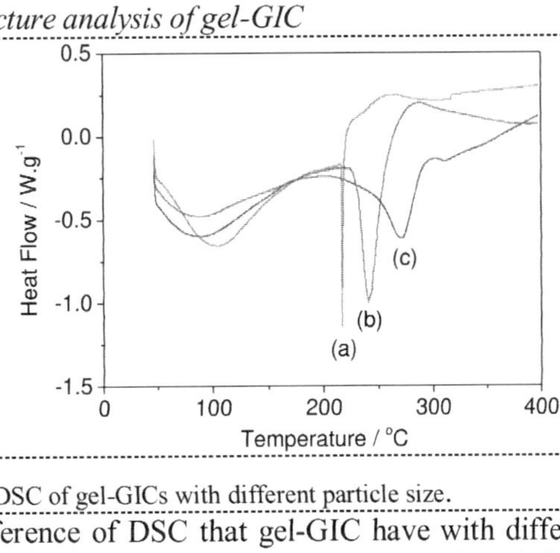

Figure 1. DSC of gel-GICs with different particle size.

The figure 1 shows the difference of DSC that gel-GIC have with different particle size. The bigger the graphite particle size is, the more heat that gel-GIC absorbed, and the higher temperature gel-GIC needs.

The SEM of the GIC (figure 2) shows the different surface state between the big gel-GIC and the small one. TiO_2 particles are all adhesived on the surface of gel-GIC. And the shapes are often spherical or spherical approximation But the particle size of gel-graphite of (a) is about 1-2μm, bigger

IOP Conf. Series: Materials Science and Engineering **761** (2020) 012012 doi:10.1088/1757-899X/761/1/012012

than the particle size of gel-graphite of (C). And also, the micro-lays of big gel-GIC are larger than small gel-GIC as pure GIC have [12]. It shows the solution-gelatin method did not affect GIC

The XRD patterns of gel-GIC are showed in figure 3. The peaks at about 28°, 55° and 87° become weak with the particle size decrease. And the diffraction intensity of sample is also weak from (a), (b) and (c), indicating that the bigger the gel-GIC is, the higher crystalline gel-GIC is.

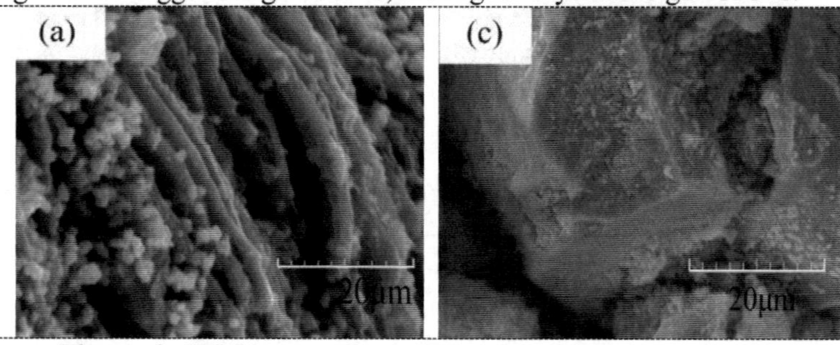

Figure 2. SEM image of gel-GICs with different particle size

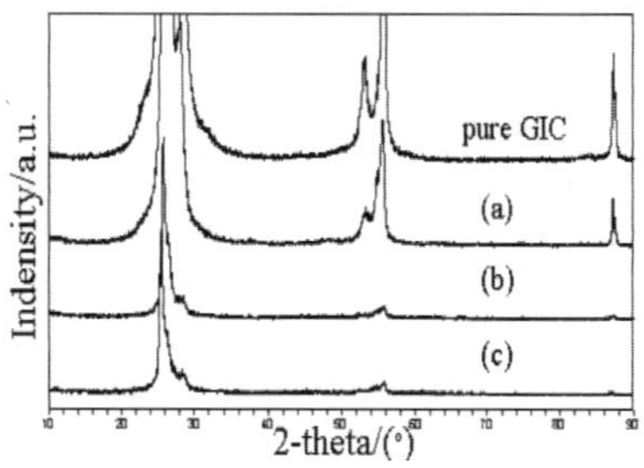

Figure 3. XRD of gel-GICs with different particle size

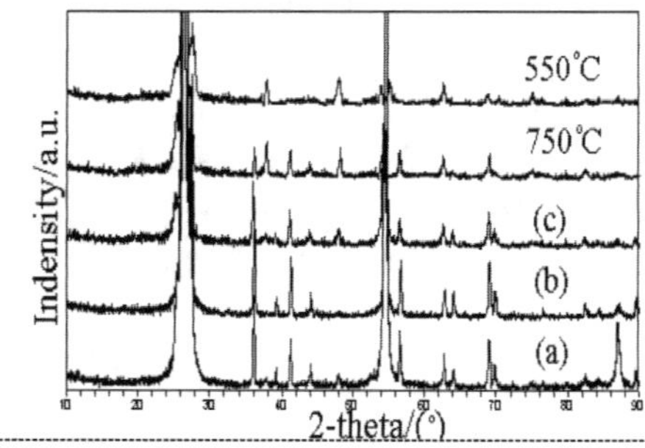

Figure 4. XRD of TiO_2/expanded graphite with different particle size

3.3. Structure and microstructure analysis of TiO2/expanded graphite

XRD patterns of prepared samples at 950°C are presented in figure 4. With the larger of the particle size, the greater the degree of rutile type. So the particle size is important for anatase phase. It also shows the temperature is the key for TiO_2. Carbon formed need to suppress the phase transformation from anatase to rutile, which usually occurred on pure TiO_2 formed by the present method above 550°C, Carbonization temperature being the higher, the amount of rutile formed was larger.

The lower temperature is necessary if anatase phase wants to be obtained. But the effect of temperature was listed in figure 5. The expansion volume is down if the temperature lower. The best condition is as follows: The particle size is more than 180μm, the temperature is between 650°C-750°C. Under this condition, the higher expansion volume (>200ml/g) and anatase phase compounds can be obtained.

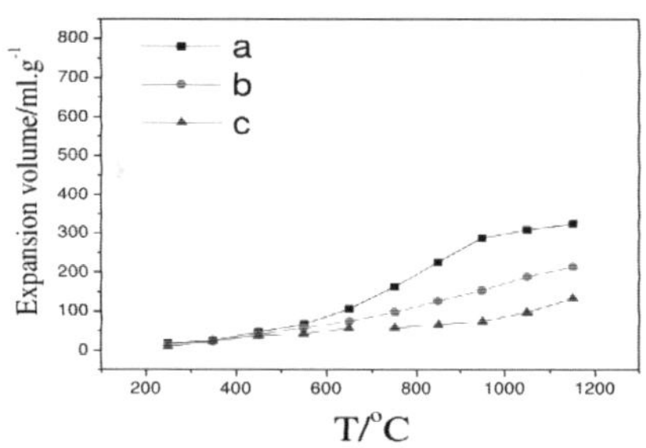

Figure 5. Expansion volume of expanded graphite

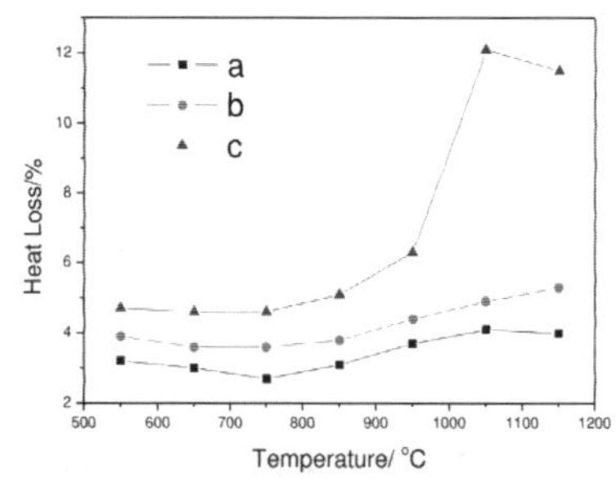

Figure 6. The heat loss at the different temperatures

at the different temperatures		

The influence of temperature lies not only on the expansion volume of graphite and crystal type of titanium dioxide, but also heat loss of compounds. The heat loss measurements at the different temperatures are presented in figure 6. As the temperature rises, there is a little downward trend with the range from 550°C to 700°C, and the heat loss is of a lower rate of not more than 5% at this stage. At the higher temperature, the heat loss of expanded graphite has a basic upward trend.

Electron spectroscopy of TiO_2/expanded graphite was shown in table 2. The contents of Ti are as follows: (a) is 1.01%, (b) is 3.60%, (c) is 1.51%. So (b) has the most element of Ti. The higher expansion volume(>200ml/g), the temperature of 550°C-650°C and high contents of anatase phase compounds can be obtained with the (b).

Table 2. EDS of TiO2/expanded graphite with different particle size

	Separation energies		
Atomic ratio	a	b	c
C	87.74	90.16	83.47
O	11.24	6.09	15.01
Ti	1.01	3.60	1.51

4. Conclusion

The better process conditions with high expansion volume are as follows: graphite that particle size is more than 180μm, expansion temperature is about 650°C-750°C. The contents of Ti in compounds of anatine TiO_2/expanded graphite can be reached 3.16%; the volume of compounds can be reached 234ml/g. The results showed that the lower temperature and smaller particle size of graphite is benefit for anatine type of TiO_2, but higher temperature is good to high expansion volume. The preparation condition should be balanced with crystal type, heat loss and expansion volume.

5. References

[1] Tai Chengan, WenBingzhang, GuiYingli JiaMofu, *et al.* Increase the Degradation Efficiency of Organic Pollutants with a Radical Scavenger （Cl^-） in a Novel Photoelectrocatalytic Reactor [J]. *Chinese Chemical Letters*, 2004, 15(4): 455-458(in Chinese).

[2] Danion Anne, Disdier Jean, Guillard Chantal et al. Photocatalytic degradation ofirnidazolinone fungicide in TiO2 -coated optical fiber reactor. *ApplCatalyB: Envionm*, 2006, 62: 274-281.

[3] Yazi Liu, Shaogui Yang, Jun Hong, Cheng Sun. Low-temperature preparation and microwave photocatalytic activity study of TiO2-mounted activated carbon. *Journal of Hazardous Materials*. 2007 (142) 208–215.

[4] Gianluca Li Puma, Awang Bono, Duduku Krishnaiah et al. Preparation of titanium dioxide photocatalyst loaded onto activated carbon support using chemical vapor deposition: A review paper. *Journal of Hazardous Materials*, 2008, 157(2-3): 209-219.

[5] Narihito Tatsuda, Hiroshi Itahara, Norihiko Setoyama, et al. Preparation of titanium dioxide/activated carbon composites using supercritical carbon dioxide. *Carbon*, 2005, 43(11): 2358-2365.

[6] Edward Carpio, Patricia Zúñiga, Silvia Ponce, et al. Photocatalytic degradation of phenol using TiO2 nanocrystals supported on activated carbon. *Journal of Molecular Catalysis A: Chemical*, 2005, 228(1-2): 293-298.

[7] Edward Carpio, Patricia Zúñiga, Silvia Ponce, et al. Photocatalytic degradation of phenol using TiO2 nanocrystals supported on activated carbon. *Journal of Molecular Catalysis A: Chemical*, 2005, 228(1-2): 293-298.

[8] Jianwen Shi, Jingtang Zheng, Peng Wu, Xiaojing Ji. Immobilization of TiO2 films on activated carbon fiber and their photocatalytic degradation properties for dye compounds with different molecular size. *Catalysis Communications*, 2008, 9(9): 1846-1850.
[9] Rusheng Yuan, Rongbo Guan, Ping Liu, et al. Photocatalytic treatment of wastewater from paper mill by TiO2 loaded on activated carbon fibers. *Colloids and Surfaces A: Physicochemical and Engineering Aspects*, 2007, 293(1-3): 80-86.
[10] M. S. Dresselhaus, G. Dresselhaus Massachusetts Institute of Technology, Cambridge, MA 02139, USA. *Advances in Physics*, 51(1) (2002), 1-186.
[11] Kong Feiyu， Yong-ping Zheng, Heng Zhao, Hai-nin Wang, Lu-ning Wang, Wan-ci Shen,et al. Sorption of heavy oils and biomedical liquids into exfoliated graphite—Research in China. *New Carbon Materials*, 18 (3) (2003), 161-173 (in Chinese).
[12] Cheng Pan,Benhua Huang,Xiaoyong Li,Hua Zhu,Donghua Zhang,Aqun Zheng,Yu Li,Yang Sun. Synthesis and Catalytic Property of Fibrous Titanium-Containing Graphite Oxide[J]. Catalysis Surveys from Asia, 2017, 21(4).

IOP Conf. Series: Materials Science and Engineering **761** (2020) 012013 doi:10.1088/1757-899X/761/1/012013

Key Structural Parameters Analysis of Effective HydroClean Drill Pipe

Chen Tao[1,2,a] Jin Yuguo[3,b] Chen Feng[3,c] Di Qinfeng[4,d] Wang Wenchang[4,e] Li Qing[1,f]

[1]School of Computer Engineering and Science, Shanghai University, Shanghai 200072, China
[2] Admissions and Career Services Office, Shanghai University, Shanghai 200072, China
[3] School of Mechatronics Engineering and Automation, Shanghai University, Shanghai 200072, China
[4] Shanghai Institute of Applied Mathematics and Mechanics, Shanghai University, Shanghai 200072, China

[a] chentao0709@shu.edu.cn, [b] 2603765066@qq.com, [c] chenfeng536@126.com, [d] qinfengd@sina.com, [e] wangwenchang1982@163.com, [f] qli@shu.edu.cn

Abstract. In directional well, horizontal well and extended reach well, there is a common problem that cuttings bed is easy to form in the process of drilling in highly deviated section. To solve this problem, an Effective Hydro Clean Drill Pipe (EHCDP) is designed, and the flow characteristics of the annular flow field under the action of EHCDP are analyzed by using computational fluid dynamics (CFD) simulation technology. On this basis, the key structural parameters affecting the swirl characteristics of EHCDP are studied. The results show that when the number of spiral grooves is 3, the axial and tangential velocities of annular fluid in spiral groove section will be greater, which is conducive to carrying and removing cuttings. Compared with the inclination angle of spiral groove with 1000 mm lead, that of spiral groove with 500 mm lead is larger, which has more obvious stirring effect on flow field, and the effect of cuttings removal is better. In addition, the spiral groove with 90 degree groove angle is more conducive to cuttings removal than the spiral groove with 30 degree groove angle. The results can provide a basis for the selection and optimization of structural parameters of spiral groove.

1. Introduction

The increasingly harsh drilling conditions and progressive drilling technology make directional drilling technology more and more widely used. Directional wells (especially directional wells with high deviation and large extended reach) in various oilfields account for an increasing proportion of their drilling. With the deepening of exploration and development, directional drilling technology has developed rapidly. At the same time, the horizontal section of horizontal wells should be extended longer, drilling speed faster and wellbore smoother, so as to develop oil and gas resources economically and efficiently [1-2]. Horizontal well drilling technology was born in Europe in the 1980s, and quickly swept across the oil drilling and production industry [3]. According to statistics, more than 50,000 horizontal wells have been drilled in more than 60 countries and regions such as the

United States, Canada and Russia in 2014, of which 88.4% are in the United States and Canada [4]. In China, horizontal well technology has been applied to varying degrees in Daqing Oilfield, Tarim Oilfield, Jidong Oilfield and Liaohe Oilfield [5]. The development of horizontal well technology has solved many technical problems for drilling industry. However, both practical and experimental studies show that because of the large deviation and horizontal displacement of such directional wells, it is difficult to carry cuttings, so cuttings are easy to accumulate and form cuttings beds during drilling.

In the drilling process, the deposition of cuttings and the formation of cuttings bed may lead to a variety of complex downhole problems. It mainly includes: 1) Testing tools and logging tools are easily blocked, causing difficulties in testing and measuring. 2) It will cause difficulties in casing and cementing operations and poor cementing, which can easily lead to oil, gas and water interflow. 3) The existence of cuttings bed may lead to the phenomenon of high friction, high torque and drag pressure of drill pipe, and even serious downhole accidents such as sticking and breaking of drilling tools [6-9]. This will affect the safety of drilling operations and drilling efficiency.

The author once investigated the cuttings bed removal technology at home and abroad, and analyzed the working mechanism of EHCDP by using CFD technology [10]. On this basis, this paper will further study the influence of key structural parameters of EHCDP spiral groove on the removal effect of cuttings bed, so as to provide basis for the selection and optimization of structural parameters.

2. Hydrodynamic model of EHCDP

The key feature of EHCDP is that the spiral groove structure is arranged, as shown in Figure 1a, and the profile of spiral groove is shown in Figure 1b. The basic structural parameters used in this paper are as follows: the outer diameter of drill pipe is 127 mm, the maximum outer diameter of spiral groove section is 165.1 mm, the borehole diameter is 215.9 mm and the corresponding eccentricity is 15 mm [10].

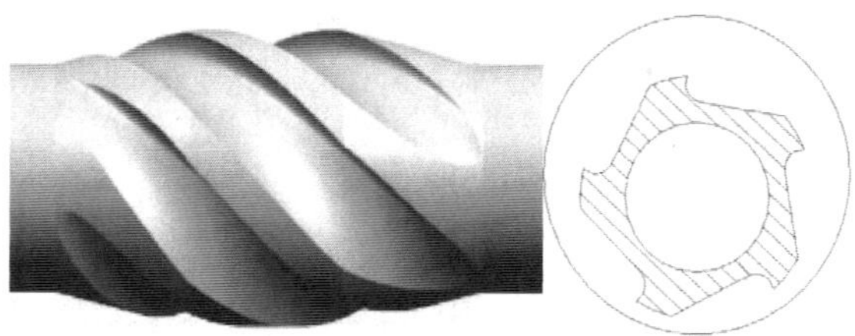

(a) Spiral groove structure (b) Spiral channel profile

Figure 1. The structure of EHCDP.

In practice, EHCDP will rotate at a certain speed (usually 60 r/min), which will cause the solid boundary to change with time. Therefore, the fluid flow in annulus must be calculated by unsteady flow. For this kind of unsteady flow problem, it is difficult to use integrated grid to calculate. Therefore, the fluid near the surface of drill pipe is divided into different computational domains by sliding mesh technology, so that the flow field control equation can be solved in each subdomain. At the interface of sub-domains, the flow field information of each sub-domain is exchanged by converting the velocity into absolute velocity. In addition, the whole fluid region is divided into hexahedral meshes, the meshes at the inner wall are shown in Figure 2a, and the meshes at the channel section are shown in Figure 2b [10]. Among them, the purple region and the green region represent different computational domains.

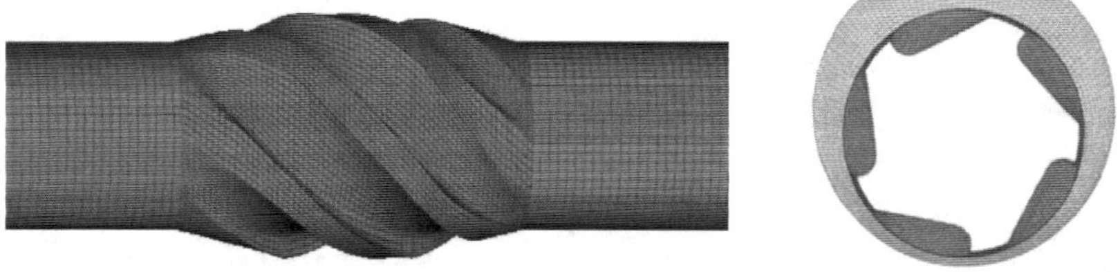

(a) Outer surface of drill pipe (b) Spiral channel profile

Figure 2. Grid diagram of spiral groove.

In this paper, the realizable k-ε turbulence model is selected for the governing equation of hydrodynamics. The transport equations of turbulent kinetic energy k and turbulent dissipation rate are as follows [11, 12]:

$$\frac{\partial(\rho k)}{\partial t}+\frac{\partial(\rho k u_i)}{\partial x_i}=\frac{\partial}{\partial x_j}\left[\left(\mu+\frac{\mu_t}{\sigma_k}\right)\frac{\partial k}{\partial x_j}\right]+G_k-\rho\varepsilon \tag{1}$$

$$\frac{\partial(\rho\varepsilon)}{\partial t}+\frac{\partial(\rho\varepsilon u_i)}{\partial x_i}=\frac{\partial}{\partial x_j}\left[\left(\mu+\frac{\mu_t}{\sigma_\varepsilon}\right)\frac{\partial\varepsilon}{\partial x_j}\right]+\rho C_1 E\varepsilon-\rho C_2\frac{\varepsilon^2}{k+\sqrt{v\varepsilon}} \tag{2}$$

In addition, the inlet of the model is calculated with a velocity boundary of 1.5m/s (that is, the flow rate of the fluid is 36 L/s). The outlet adopts pressure boundary with the size of 30 MPa. The borehole wall is fixed. And the outer surface of the drill pipe is a rotating wall with a rotating speed of 60 r/min [10].

3. Analysis of the effect of key structural parameters of EHCDP on the flow characteristics of annular flow field

First of all, it should be noted that: Lower static pressure in the middle of spiral groove area, higher dynamic pressure in small annulus, greater axial and tangential velocities will be more conducive to carrying cuttings and clearing cuttings bed. The swirl structure generated by spiral groove can also improve the removal effect of cuttings.

3.1. Analysis of the influence of channel number on annular flow field

Figure 3 shows the hydrostatic pressure distribution in annulus under different number of helical groove structures. It can be seen from the graph that they have similar static pressure distributions for annular fluids. The high pressure zone is formed near the inlet side of each groove, and the low pressure zone is formed near the hook angle of each spiral groove. However, with the increase of the number of grooves, the hydrostatic pressure of annular fluid in spiral groove section increases gradually. This makes it more difficult for the cuttings to enter the hook angle area, which is not conducive to the removal of cuttings.

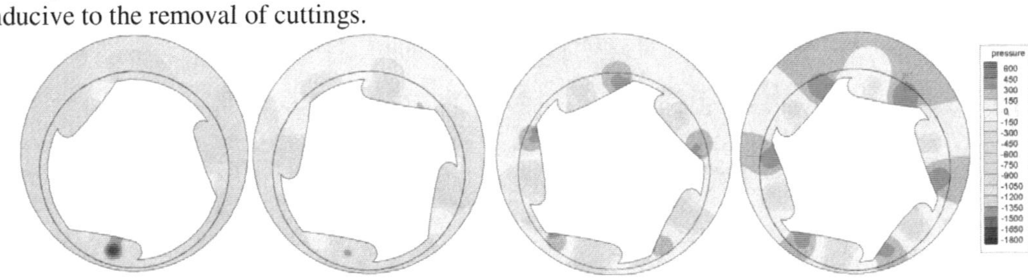

Figure 3. Distribution of hydrostatic pressure in annulus.

Figure 4 shows the hydrodynamic pressure distribution in annulus under different number of helical groove structures. From the graph, it can be seen that the dynamic pressure distribution law is similar. They all have high dynamic pressure in small annulus. At the same time, with the increase of the number of spiral grooves, the dynamic pressure of annular fluid in spiral groove section decreases gradually, including small annulus. This is because with the increase of the number of spiral grooves, the annulus space will gradually increase, resulting in a gradual decreaseof dynamic pressure. However, the decrease of dynamic pressure in small annulus is not conducive to the erosion of bottom hole cuttings.

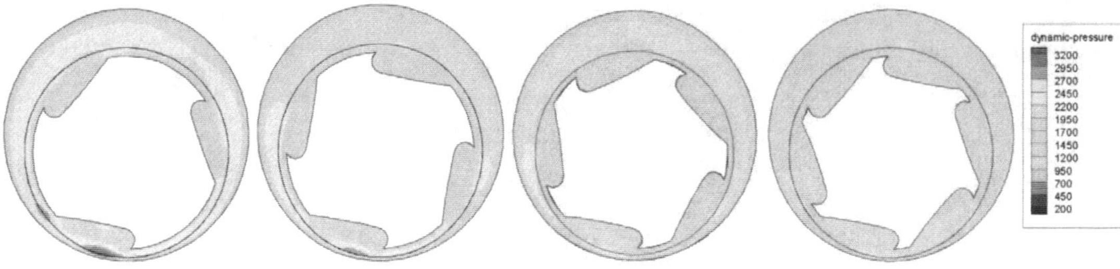

Figure 4. Distribution of hydrodynamic pressure in annulus.

Figure 5 shows the tangential velocity vector of annular fluid under the action of different number of spiral groove structures. From the graph, it can be seen that the annular fluid has a higher positive tangential velocity in the small annulus under the action of various spiral groove structures, and a higher reverse tangential velocity in the center of each spiral groove. At the same time, according to the direction of velocity vectors in the graph, they all produce vortices which are beneficial to cuttings removal. This also shows the influence of spiral channel structure on annular flow field and the feasibility of using spiral channel structure to help remove cuttings bed. However, as the number of spiral grooves increases, the tangential velocity of annular fluid in spiral groove section decreases gradually.

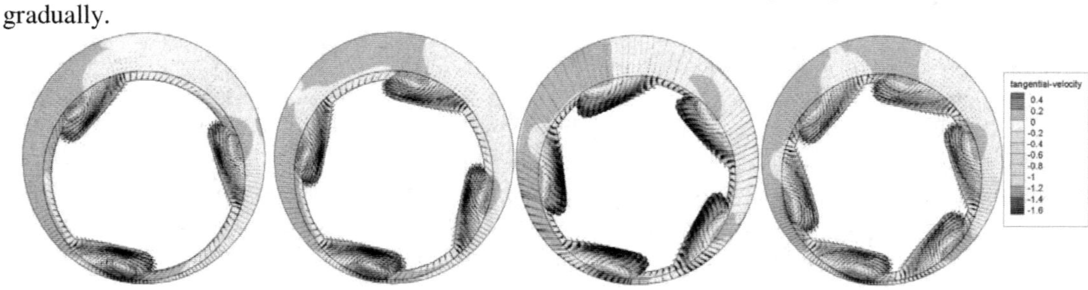

Figure 5. Tangential velocity vector diagram of annular fluid.

Figure 6 shows the axial velocity distribution of annular fluid under different number of helical groove structures. As can be seen from the graph, they all have higher axial velocity in the small annulus. Accordingly, with the increase of the number of spiral grooves, the axial velocity of annular fluid in spiral groove section decreases. The small axial velocity will not be conducive to carrying cuttings out from the bottom of the well.

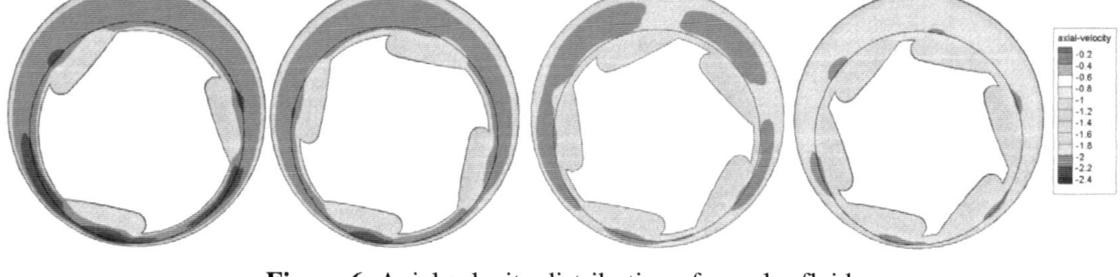

Figure 6. Axial velocity distribution of annular fluid.

Figure 7 shows the inner streamline of annulus fluid at z = 0.5m (the beginning of the spiral channel), 0.625m, 0.75m (the end of the spiral channel), 0.875m, 1m, 1.125 m, 1.25m and 1.375m respectively. It can be seen from the graph that, in general, with the increase of the number of grooves, the influence range of eddy current structure generated by spiral grooves becomes smaller. In terms of eddy current characteristics, it is also not conducive to cuttings removal.

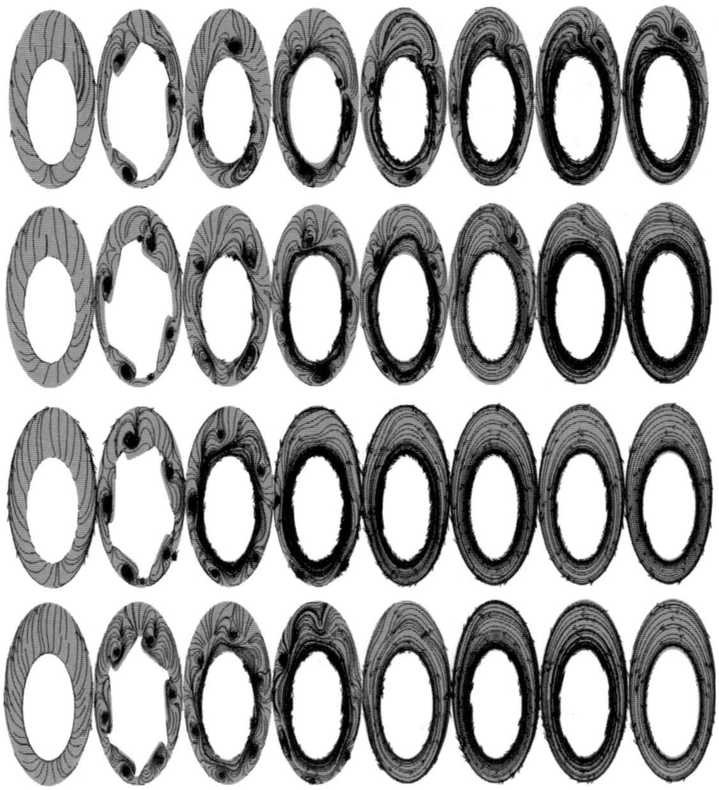

Figure 7. Streamline diagrams of annular fluids.

To sum up, from the analysis results of static pressure, dynamic pressure, axial velocity and tangential velocity, it can be concluded that when the number of spiral grooves is 3, 4, 5 and 6 respectively, the spiral groove structure with the number of 3 has the best cleaning effect on cuttings bed.

3.2. Analysis of the influence of channel inclination on annular flow field

The inclination angle of the spiral groove will have an important influence on the cleaning effect of the spiral groove. In this paper, the inclination of spiral groove is changed by changing the parameter of lead. And the cleaning effect of spiral groove with different lead is also studied. Figure 8 is a schematic diagram of spiral grooves with different lead.

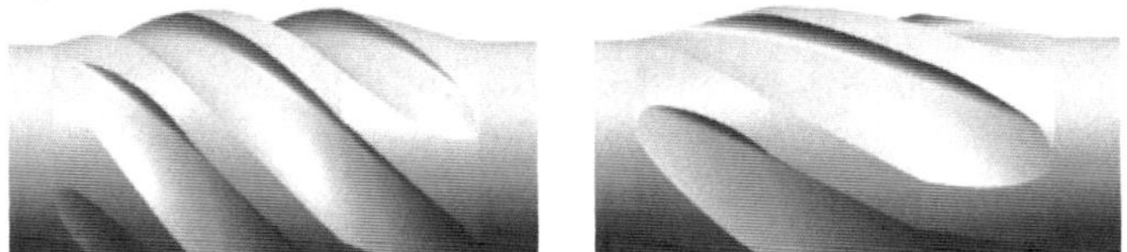

(a) Lead = 500 mm (b) Lead = 1000mm

Figure 8. Schematic diagram of spiral groove with different lead.

Figure 9 and Figure 10 show that the hydrostatic and hydrodynamic profiles of annulus fluids are z=0.5m (at the beginning of the spiral channel), 0.625m, 0.75m (at the end of the spiral channel), 0.875m, 1m, 1.125 m, 1.25m and 1.375m respectively.

(a) Lead = 500 mm

(b) Lead = 1000mm

Figure 9. Static pressure distribution of annular fluids.

(a) Lead = 500 mm

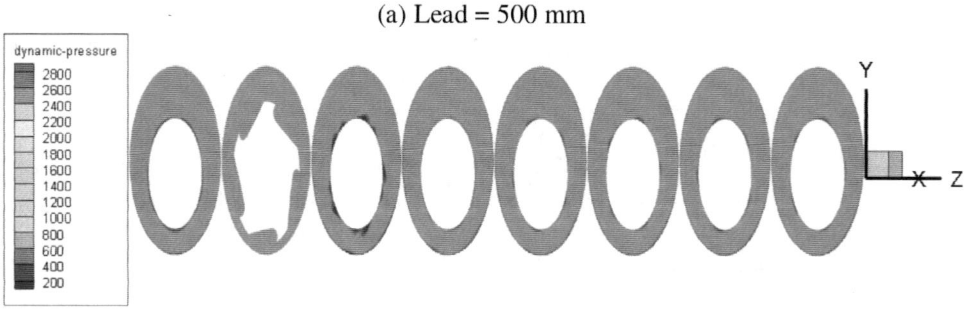

(b) Lead = 1000mm

Figure 10. Dynamic pressure distribution of annular fluids.

It can be seen that the fluctuation of hydrostatic and dynamic pressure in the radial direction of annulus fluid is relatively smaller when the lead of spiral groove is 1000mm. Not only in the spiral groove area and its vicinity, but also in many sections far from the groove. That is to say, when the lead of spiral groove is 1000mm, its stirring effect on annular fluid is relatively weak, and its cleaning effect is even weaker. The pressure distribution at the spiral groove can also be seen that when the lead is 1000mm, it is not conducive to the removal of cuttings compared with the 500 mm lead. Similar conclusions can be drawn from the analysis of Figure 11 and Figure 12. These two diagrams are the distributions of the axial and tangential velocities of annular fluid in the longitudinal section at x=0 under the action of helical groove structures with different lead. As shown in Figure 11, the overall axial velocity of annulus fluid decreases with the change of lead from 500 mm to 1000 mm, which will be detrimental to the carrying of cuttings. As can be seen from Figure 12, the larger the lead, the

IOP Conf. Series: Materials Science and Engineering **761** (2020) 012013 doi:10.1088/1757-899X/761/1/012013

smaller the tangential velocity of the annular fluid. In particular, the decrease of tangential velocity in small annulus will not be conducive to the substitution of cuttings into large annulus and its removal.

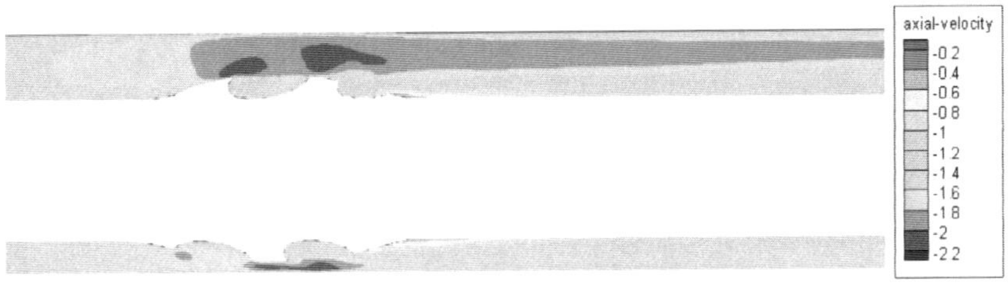

(a) Lead = 500 mm

(b) Lead = 1000mm

Figure 11. Axial velocity distribution of annular fluid.

(a) Lead = 500 mm

(b) Lead = 1000mm

Figure 12. Tangential velocity distribution of annular fluid.

From the above analysis data, it can be concluded that the spiral groove structure with 500 mm lead is more advantageous to the removal of cuttings bed than the spiral groove structure with 1000 mm lead.

3.3. Analysis of the influence of groove angle on annular flow field

In this paper, the cleaning effect of spiral groove with groove angles of 30 and 90 respectively is studied. Figure 13 is a schematic diagram of the groove angle of the spiral groove.

IOP Conf. Series: Materials Science and Engineering **761** (2020) 012013 doi:10.1088/1757-899X/761/1/012013

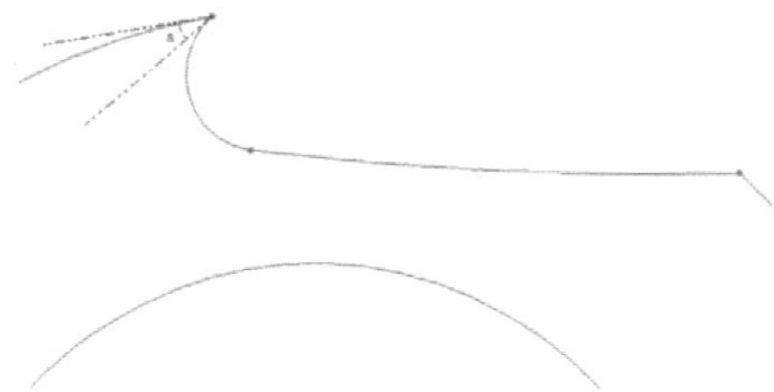

Figure 13. Drawing of groove angle

From Figure 14 and 15, it can be seen that compared with spiral grooves with 30 degrees groove angle, spiral grooves with 90 degrees groove angle will make the annular flow field have lower static pressure and higher dynamic pressure. As mentioned above, this will be more conducive to cuttings removal.

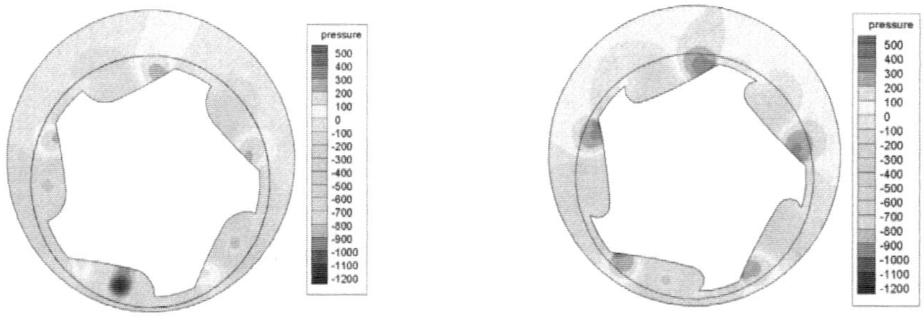

(a) 90 degrees (b) 30 degrees

Figure 14. Static pressure comparison of annular flow field

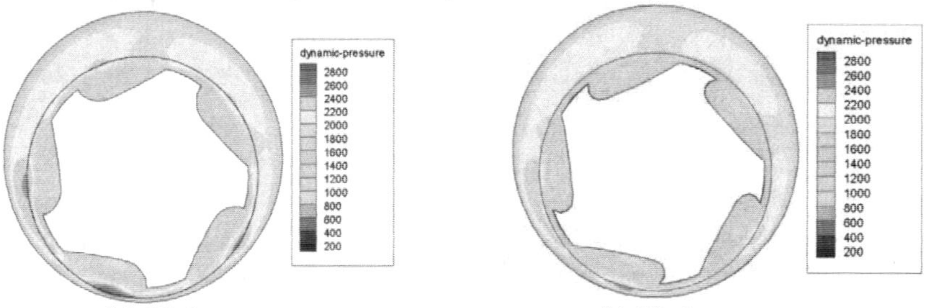

(a) 90 degrees (b) 30 degrees

Figure 15. Dynamic pressure comparison of annular flow field

From Figure 16 and 17, it can be seen that the spiral groove with 90 degree groove angle will make the annular flow field have higher axial and tangential velocities than the spiral groove with 30 degree groove angle, which is also more conducive to cuttings removal.

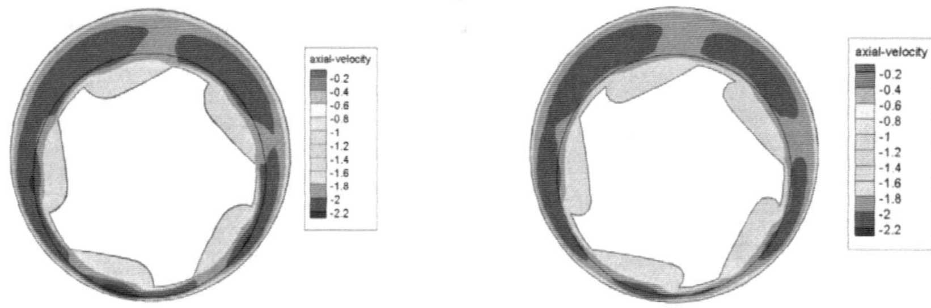

(a) 90 degrees (b) 30 degrees

Figure 16. Axial velocity comparison of annular flow field

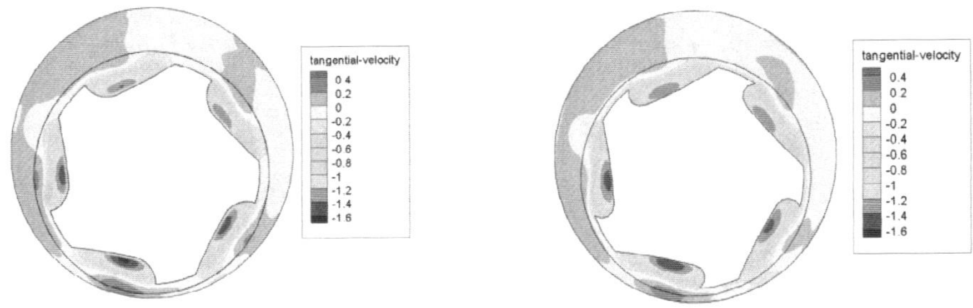

(a) 90 degrees (b) 30 degrees

Figure 17. Tangential velocity comparison of annular flow field

In conclusion, compared with the spiral groove with 30 degree groove angle, the spiral groove with 90 degree groove angle will make the annular flow field have better cleaning effect.

4. Conclusions

Through the numerical analysis of the annular flow field under the action of EHCDP, the following conclusions can be drawn:

1) Under the action of EHCDP, a strong swirl appears in the annular flow field, and a strong eddy appears near the spiral channel, which is beneficial to the removal of cuttings.

2) For drill pipes with spiral grooves of 3, 4, 5 and 6, the smaller the number of grooves, the greater the fluctuation of annular flow field. Moreover, all kinds of pressure and velocity analysis are more conducive to improving the removal effect of cuttings. That is to say, when the number of channel is 3, its cuttings removal effect is the best.

3) The spiral groove with 500 mm lead has a larger inclination angle than that with 1000 mm lead, which has a more obvious stirring effect on the flow field and better cuttings removal effect.

4) The spiral groove with 90 degree groove angle will make the annular flow field more conducive to cuttings removal than the spiral groove with 30 degree groove angle.

5. References

[1] Wang Minsheng, Guang Xinjun. Advances and Trend of Directional Drilling Technology[J]. CHINA PETROLEUM MACHINERY, 2015, 43 (7).

[2] Wang Jialu, Horizontal well drilling technology and its application in petroleum development[J], Chemical Enterprise Management, 2014.

[3] Jafei, Wang Yongning, Qu Haipeng, et al. Countermeasure analysis of horizontal well technology in oilfield production management[J]. Chemical Enterprise Management, 2018(11).

[4] Tan Heping, Research on drilling technology for extended reach horizontal wells[J], China Petroleum and Chemical Standard and Quality, 2014.

[5] Jiyu. Review of the Development of Horizontal Well Track[J]. WORLD WELL LOGGING TECHNOLOGY, 2016 (05): 5+27-30.

[6] Sun Yongxing. Weiyuan-Changning area shale gas long horizontal section horizontal well anti-sticking countermeasures[A]. China Petroleum Association Natural Gas Commission, SICHUAN PETROLEUM INSTITUTE, ZHEJIANG PETROLEUM INSTITUTE. Annual National Natural Gas Academic Conference in 2017[C]. China Petroleum Association Natural Gas Commission, SICHUAN PETROLEUM INSTITUTE, ZHEJIANG PETROLEUM INSTITUTE: China Petroleum Association Natural Gas Commission, 2017:8.

[7] Qu Jingyu. Numerical Simulation of Flow Field Analysis of Hole Cleaning Tool Based on TFM [D]. Northeast Petroleum University, 2018.

[8] Xiao Xujun. Research on Drilling Reduction and Anti-lock of Horizontal Well [J]. Chemical Engineering Design Communications, 2017 (8).

[9] Wei Wenzhong, Liu Yongwang, Guan Zhichuan, et al. A new device for cleaning cuttings bed in large inclination and extension wells[J]. Journal of Oil and Gas Technology, 2006, 28 (5): 138-140.
[10] Chen Feng, Diqinfeng, Yuan Pengbin, et al. Mechanism of an effective hydroclean drill pipe for hole cleaning. ACTA PETROLEI SINICA, 2012 (02): P. 298-303.
[11] Shih T H , Liou W W , Shabbir A , et al. A new eddy viscosity model for high reynolds number turbulent flows[J]. Computers & Fluids, 1995, 24(3): 227-238.
[12] Wang Fujun. Analysis of computational fluid dynamics: theory and applications of CFD software[M]. Beijing: Tsinghua University Press, 2004:125-126.

Acknowledgements
This research was supported in part by the National Science Funding of China (U1663205, 51704191, 51804194), Shanghai Leading Academic Discipline Project (S30106), the Shanghai Municipal Education Commission (Peak Discipline Construction Program), and the Shanghai Sailing Program(17YF1428000).

IOP Conf. Series: Materials Science and Engineering **761** (2020) 012014 doi:10.1088/1757-899X/761/1/012014

Theoretical Analysis of Two Evaporator Configurations for a Conventional Refrigerator Coupled To a Phase Change Material

M Berdja[1], F Yahi[1], A Tetbirt[1], M Ouali[1], M A Djebiret[1], M Mokrane[1]
[1] Unité de Développement des Equipements Solaires. UDES./ Centre de Développement des Energies Renouvelables. CDER. Bou-Ismail.42415.w.Tipaza. Algeria.
E-mail: berdjamohand@gmail.com / mohand.berdja@udes.dz

Abstract. In this paper, we present the study and development of prototypes of refrigeration systems that operate more efficiently and consume less electrical energy, and this based on the use of Phase change materials (PCM) as thermal energy storage (TES). These materials have the ability to store thermal energy on its latent heat form, which is produced during off-peak periods, and then release this thermal energy to the consumer during the desired period or during periods when electrical demand is high. The effect of phase change material on the energy performance of a conventional refrigerator has been highlighted for two different evaporator configurations; a flat evaporator and a ventilated finned evaporator. An analysis to select the adequate phase change material (PCM) as well as its arrangement to be integrated into the refrigerator, and the mechanism by which the PCM affects the refrigerator operating conditions according to the type of evaporator has also been discussed.

1. Introduction

The overall electricity consumption continues to increase from year to year, which encourages production companies to operate frequent load shedding so that the instantaneous and non-storable electricity production can meet consumer demand. This increase in overall consumption is mainly due to the increase in population, development of living standards and access to technologies, and also to the excessive and non-rational use of electrical equipment especially the refrigerating appliances. Therefore, It is more than necessary to design refrigerators systems that operate more efficiently (higher COP), with longer downtime and low energy consumption. The use of thermal storage with phase change materials (PCM) seems to be one of the least expensive solutions with no environmental impact [1-3].

A PCM is defined as any substance that has a significant latent heat that occurs at a certain temperature. This temperature is constant for pure and eutectic materials, and may represent a range of few degrees for mixtures and binary solutions [4]. Only the liquid-solid phase change to a practical interest in the field of thermal energy storage, even if the liquid-gas phase change may involve latent heat more important than that of the liquid-solid transformation. These materials have the advantage to offset the gap between the production of thermal energy (cold /heat) and the needs and to have a high energy density compared to a sensible storage. Nevertheless, PCM can present some disadvantages, such as low thermal conductivity and latent heat, supercooling phenomena, high price, corrosiveness and toxicity [5].

The use of PCMs as a thermal storage device concerns several fields of application. It is found mainly in the field of building, solar, and also in refrigeration and air conditioning. Regarding the application of PCMs in refrigeration, the PCM can be integrated in the refrigerator in different ways: In contact with refrigerator air [6], in contact with evaporator [7], in contact with condenser [8] and evaporator immerged in PCM [9].

In this study, two (2) configurations of domestic refrigerators are considered; one supplied with a flat evaporator and with a ventilated finned evaporator, in order to highlight the effect of the phase change material (PCM) on the refrigerator operating conditions and its thermal performance.

2. Effect of PCM on the refrigerator operating conditions

2.1. Flat plate evaporator

The presence of a PCM layer at the evaporator's surface (Figure 1.a) permits an extension of the compressor's downtime. The cold latent heat stored into PCM during the compressor's on cycle is released partly to the air, and partly to the evaporator's surface. The temperature of the thermostat bulb remains at a low temperature, allowing the compressor to be idle for a longer time.

The PCM also improves heat transfer at the evaporator's surface, and permits an easier evaporation of the refrigerant and thus an improvement of the evaporator performance. This results in a slight increase of the evaporation temperature and the COP.

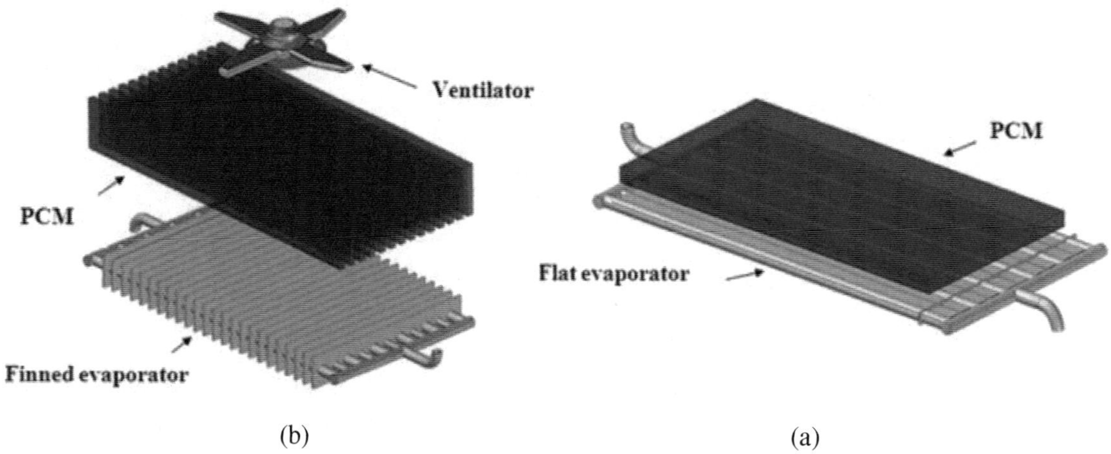

Figure 1. (a) Flat plate evaporator coupled to PCM. (b) Finned ventilated evaporator coupled to PCM.

However, thicker layers of the PCM could allow for additional thermal resistance that reduces the heat transfer that occurs between the refrigerant and the air. Thus, a thicker layer of PCM would prolong the refrigerator's downtime without being able to maintain a low air temperature inside the refrigerator (figure 2). The search for the optimum thickness of the PCM is necessary to reduce the excessive rise in air temperature inside the refrigerator. This can be done by an optimization that leads to obtain the maximum value of the compressor downtime equivalent to the minimum value of the air temperature rise in the refrigerator.

2.2. Finned evaporator

For a finned evaporator type (Figure .1.b), the configuration of a heat exchanger should be selected in order to contain the quantity of PCM to be integrated into the refrigerator. In that case, the design of the PCM heat exchanger is required to ensure an optimal overall operation and a prolonged downtime of the refrigerator. The heat exchanger could take several configurations; plate, honeycomb or packed bed. The presence of fins on the evaporator does not allow placing the PCM on the surface of the evaporator in this type of configuration.

It would be necessary to have a melting time that is equivalent to the desired compressor shutdown time. The melting time is defined as the time for which the melting of the entire mass of the PCM in the heat exchanger takes part (Figure .3.a). The melting time, which is the most important parameter for the dimensioning of the heat exchanger containing the PCM, allows a rigorous control of the compressor's operating time.

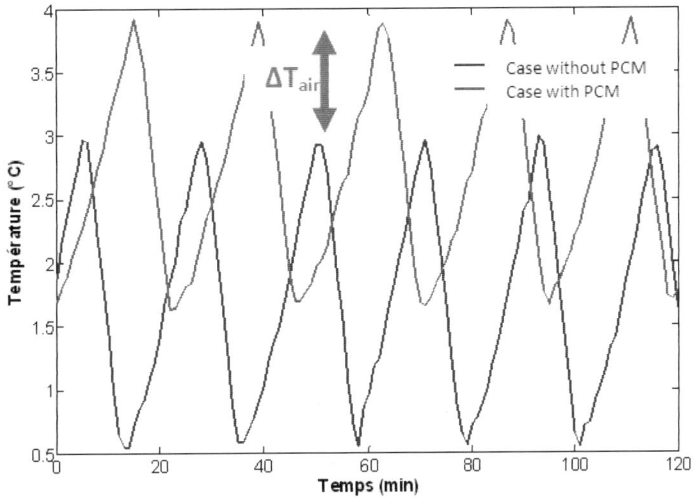

Figure 2. Air temperature profile. Case with vs without PCM.

The melting time is calculated from the following parameters:

- Thermal characteristics of PCM.
- PCM surface exchanger.
- Thermal balance through the walls of the refrigerator.
- Air velocity generated by the ventilator through the PCM heat exchanger.

However, a long shutdown requires a high amount of PCM. A large volume of PCM inside the refrigerator could limit the volume intended for the storage of foodstuffs. Beside disadvantage of the limitation of the volume due to the large the amount of PCM, melting time is also conditioned by two (2) constraints that must be minimized:

- Pressure drop at the PCM heat exchanger (Figure .3.b): a significant pressure drop through the PCM exchanger would require a sufficiently powerful fan (energy-consuming and expensive) to satisfy the air circulation inside the refrigerator, and overcoming the pressure drop. The pressure drop depends mainly on the configuration of the PCM heat exchanger (length, diameter and air velocity).
- Ventilation efficiency: the fan must be efficient energetically in its range of velocity. The minimum ventilation performance is dictated by the standard eco-design requirements for ventilation units.

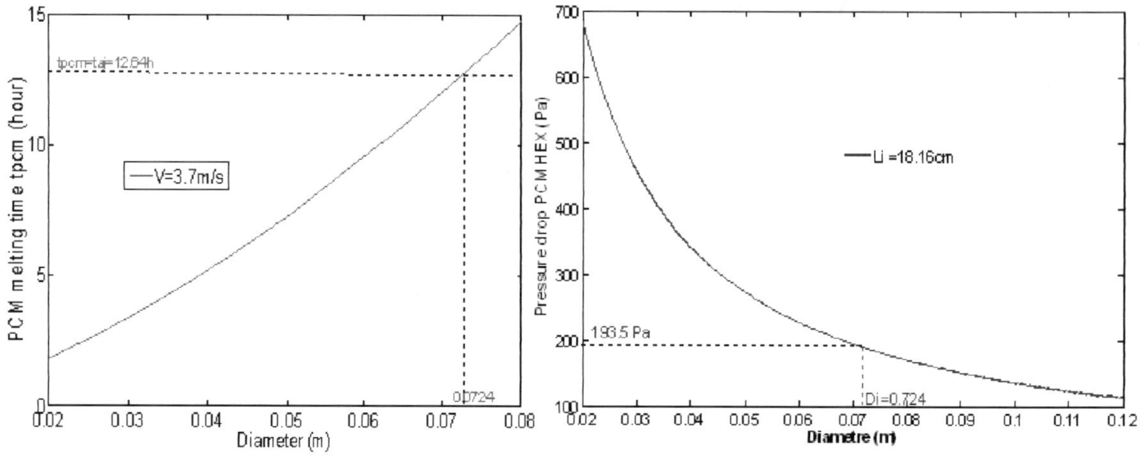

(a) (b)

Figure 3. (a) PCM melting time profile. (b) Pressure drop through PCM heat exchanger.

It would be necessary to carry out an optimization of the dimensions of PCM heat exchanger in order to limit the pressure drop through the PCM heat exchanger and allow a more efficient operating condition of the ventilator. This permits to obtain an extended compressor downtime and a save in refrigerator energy consumption.

3. Selecting the phase change temperature of the PCM

3.1. Flat plate evaporator

The definition of the phase change temperature of the PCM to be integrated in a domestic refrigerator depends on the way it is placed inside the refrigerator. The PCM interacts thermally with its surrounding environment. In this case, the PCM is in direct contact with the evaporator, and tends towards thermal equilibrium by conduction with the surface of the evaporator. The temperature of the chosen PCM must in this case be between the maximum and minimum temperature of the surface of the evaporator (Figure 4), so that the PCM can solidify during the operating cycle of the compressor.

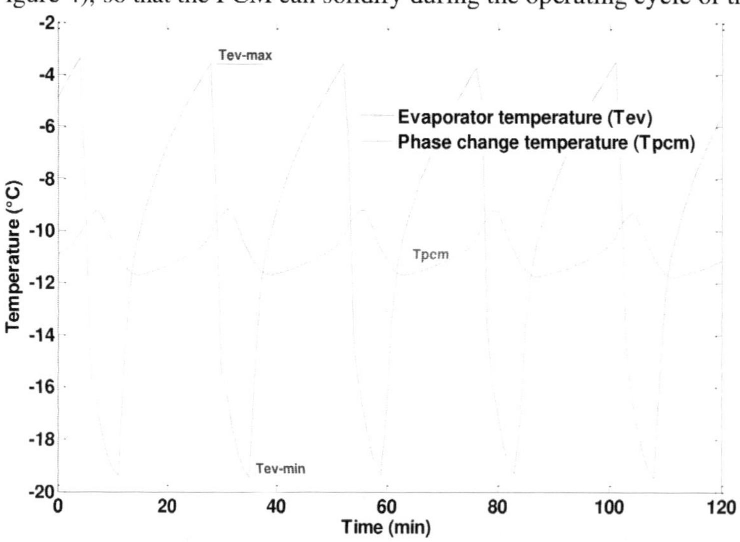

Figure 4. Evaporator temperature vs phase change temperature of PCM $T_{ev\text{-}min}<T_{PCM}<T_{ev\text{-}max}$.

3.2. Finned ventilated evaporator

In the case of a refrigerator equipped with a finned ventilated evaporator, the phase change temperature of the selected PCM should be between the maximum and minimum temperature of the air in the chamber. The PCM interacts thermally with its surrounding air, and tends towards thermal equilibrium by convection with air in the chamber. This is because the PCM is only in contact with the air, unlike the case of a plate evaporator where the PCM is in contact with the surface of the evaporator (Figure 5). The air temperature inside the refrigerator must be set to several set temperatures to find the proper temperature that would situate the phase change temperature between the maximum and minimum air temperature, and thus allow the phase change to occur.

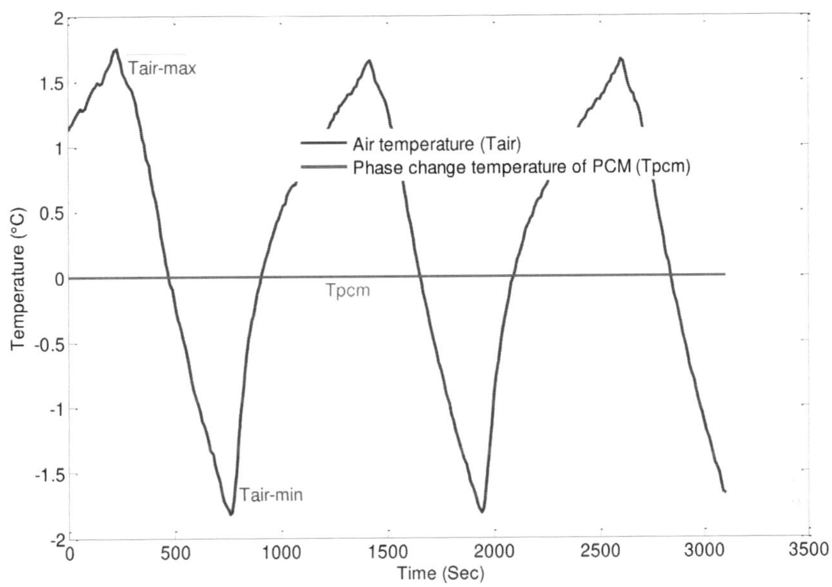

Figure 5. Air and PCM temperature (Tconsigne=-1°C). $T_{air\text{-}min} < T_{PCM} < T_{air\text{-}max}$

Manipulating the set point temperature of the thermostat makes may modify the air temperature range of the refrigerator (minimum-maximum temperature). This temperature range could shift with respect to the value of the phase change temperature of the PCM. The problem does not arise for a flat evaporator, because the difference between the maximum and minimum temperature of the surface of the evaporator is very large (around 20°C) and the manipulation of the thermostat does not allow a significant shift with respect to the phase change temperature of the PCM. On the other hand, for a finned ventilated evaporator, the difference between the maximum and the minimum temperature of the air is generally very small (2 to 4°C), and this interval shifts greatly with the manipulation of the thermostat, generating a shift with respect to the phase change temperature of the PCM (Figure 6.a and Figure 6.b).

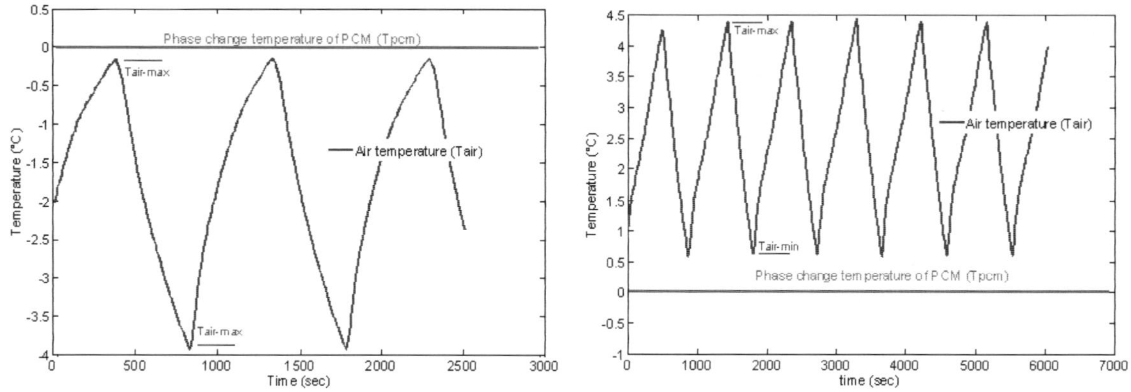

(a) (b)

Figure 6. Air and PCM temperature. (a) Tconsigne=-3.5°C. $T_{PCM}>(T_{air-max},T_{air-min})$. (b) Tconsigne= +1°C. $T_{PCM}>(T_{air-max},T_{air-min})$

4. Conclusion

An energy analysis of refrigerator using a phase change material (PCM) as a cold thermal storage device has been conducted for two (2) types of evaporators commonly used in conventional refrigeration; a flat evaporator and a ventilated finned evaporator.

The analysis to select the adequate phase change material (PCM) as well as its arrangement to be integrated into the refrigerator showed that for a flat plate evaporator, the PCM is in contact with air and evaporator surface. The PCM phase change temperature should be selected to be close to the refrigerant phase change temperature. However, for a ventilated finned evaporator, the PCM is only in contact with the room air, and the PCM temperature should be selected to be between the maximum and minimum air temperature inside the refrigerator.

The mechanism by which the PCM affects the refrigerator operating conditions differs according to the type of evaporator. For a flat evaporator without forced ventilation, The PCM improves heat transfer at the evaporator which increases the evaporator performance and extends the compressor downtime. However, beyond a certain thickness, the PCM can act as thermal insulation to the heat transfer that occurs between refrigerant and air, and increase the air temperature inside refrigerator. Finding an optimal thickness of the PCM slab is necessary for this kind of application, which is defined by a maximum of extension of the compressor downtime that corresponds to the minimum of the air temperature rise inside the refrigerator. For a ventilated finned evaporator, The PCM does not affect directly the performance of the refrigerator, but allows a more precise control of the compressor downtime. In fact, the compressor downtime is equivalent to the PCM melting time, which depends mainly on the mass of PCM contained into the heat exchanger, and the quality of the heat transfer through, but limited by an elevated pressure drop and low ventilation efficiency. An optimization of the dimensions of the heat exchanger is necessary to limit the pressure drop, to allow the reduction of the ventilation energy consumption and the extension of the compressor downtime.

5. References

[1] Desmons, J, *Aide mémoire génie climatique*, 2010, 2eme édition DUNOD.
[2] Chidambaram LA, Ramana AS, Kamaraj G and Velraj R, 2011, *Review of solar cooling methods and thermal storage options. Renewable and Sustainable Energy Reviews*, **15**, 3220-3228.
[3] Laidi, M, Hanini S, Abbad B, Berdja M and Ouali M, 2012, *The study and performance of a modified ENIEM conventional refrigerator to serve as a photovoltaic powered one under Algerian climate conditions*, Journal of Renewable and Sustainable Energy **4**, 053112.

[4] Farid M M, Khudhair A M, Razack S A and Al-Hallaj S, 2004, *A review on phase change energy storage: materials and applications*, Energy Conversion and Management **45**, 1597–1615.

[5] Zalba B, Marin J, Cabeza L F and Mehling H, 2003, *Review on thermal energy storage with phase change: materials, heat transfer analysis and applications*, Applied Thermal Engineering **23**, 251–283.

[6] Oro E, Miro L, Farid M M and Cabeza L F, 2012, *Improving thermal performance of freezers using phase change materials*, International Journal of Refrigeration **35**, 984–991.

[7] Azzouz K, Leducq D and Gobin D, 2008, *Performance enhancement of a household refrigerator by addition of latent heat storage*, International Journal of Refrigeration **31**, 892–901.

[8] Wang F, Maidment G, Missenden J and Tozer R, 2007, *The novel use of phase change materials in refrigeration plant. Part 1: Experimental investigation*, Applied Thermal Engineering **27**, 2893–2901.

[9] Berdja M, Abbad B, Laidi M, Yahi F and Ouali M, 2012, *Numerical Simulation Of A Phase Change Material (PCM) In A Domestic Refrigerator Powered By Photovoltaic Energy*, ICHMT Digital Library Online.

IOP Conf. Series: Materials Science and Engineering **761** (2020) 012015 doi:10.1088/1757-899X/761/1/012015

The Misfit Strain Energy Collapse during Precipitation in Solid Transformations

Liu Lei, Lai Qi, Li Liang and Liu Qiaochu
College of Vanadium and Titanium, Panzhihua University, Panzhihua 617000, China

Abstract. The misfit strain energy collapse (MSEC) which results in the dendritic or split γ' precipitating particles during solid transformations in Ni-base alloys is discovered. Three parameters of the MSEC function, the MSEC delay, the MSEC rate, and the initial value of the MSEC function, are correlated to the persistence of the isothermal cooling, the rate during the continuous cooling, and the initial misfit strain during nucleation respectively. The MSEC point is the intersection of the two crystal lattice parameters $a_{\gamma'}(T)$ and $a_{\gamma}(T)$ of the two different phases.

1. Introduction

The microstructural misfit strain energy theory derived by A. G. Khachaturyan [1-2] reveals that the elastic strain energy of the microstructure depends on the misfit strain between the precipitates and the matrix during phase transformations in solids, but it does not point out when and how the elastic strain energy takes place. Considering the temperature is successively changed in different stage of aging during continuous cooling, the phenomenon of misfit strain energy collapse (MSEC) might occur accompanying with the variation of the structure during precipitation. The so-called MSEC refers to the opinion that the misfit strain energy varies successively during aging but not suddenly at beginning of nucleation.

2. The temperature dependent elastic strain energy

The temperature dependent total elastic energy may be expressed by

$$E^{el} = \int_V C_{ijkl}(c,T)\varepsilon_{ij}^{el}(r,T)\varepsilon_{kl}^{el}(r,T)d^3r. \quad (1)$$

If the applied strain is not taken into account, which means that $\varepsilon_{kl}^{el} = 0$, and the elastic constant is considered to be homogeneous, i.e. $\Delta C_{ijkl}^{c}(T) = 0$, the temperature dependent total elastic energy of the aging system may then be obtained

$$E^{el} = \int_V B(\boldsymbol{n},T)|\Delta c(\boldsymbol{g},t)|^2 \frac{d^3r}{(2\pi)^3}. \quad (2)$$

The inhomogeneous part of $B(n,T)$in Eq. (2) may be given by

$$\Delta B(\boldsymbol{n},T) = B(\boldsymbol{n},T) - B(\boldsymbol{0},T) = \frac{4(C_{11}^T+2C_{12}^T)^2 n_1^2 n_2^2 \Delta}{C_{11}^T(C_{11}^T+C_{12}^T+2C_{44}^T} [\varepsilon_0^c(T)]^2. \quad (3)$$

With the definition

$$\Delta E^{el} = \frac{1}{2}\int [B(n,T) - B(0,T)]|\Delta c(g,t)|^2 \frac{d^3g}{(2\pi)^3} \quad (4)$$

and substituting Eq. (3) into Eq. (2) and combining Eq. (1), the simplified inhomogeneous part of the temperature dependent total misfit strain energy may then be obtained..

3. Temperature dependent elastic strain energy for modelled Ni-Ai-Si alloys

With the linear assumption the lattice constants of γ and γ' in the modeled Ni-base alloys may be given by

$$a_{\gamma,\gamma'} = a_0 + a_1 T + a_2 T^2, \tag{5}$$

where $a_i = (1-x)a_i^{Ni-Al} + x a_i^{Ni-Si}$. The coefficients a_i are listed in table 1.

Table 1. The coefficient a_i of crystal lattice parameters of γ' and γ in modeled Ni-base alloy abiding by the linear rule.

		Ni-17.7at.%Al		Ni-12.4at.%Al-.54at.%Si		Ni-18.47at.%Si	
		γ'	γ	γ'	γ	γ'	γ
a_0	(nm)	0.35628	0.35459	0.35467	0.35372	0.35091	0.35170
$a_1 \times 10^6$	(nm/K)	6.16199	5.74055	5.72880	5.46490	4.71791	4.82175
$a_2 \times 10^{10}$	(nm/K^2)	-11.32198	-1.00986	-5.65040	2.58670	7.58328	10.97854

The inhomogeneity of the temperature dependent misfit strain energy $\Delta E^{el}(T)$ is given in Eq. (3), in which the temperature dependent misfit strain $\varepsilon_0^c(T)$ is a function of the temperature. With the assumption of the homogeneous temperature, the heat treatment regime may be simplified by function $T(t)$.

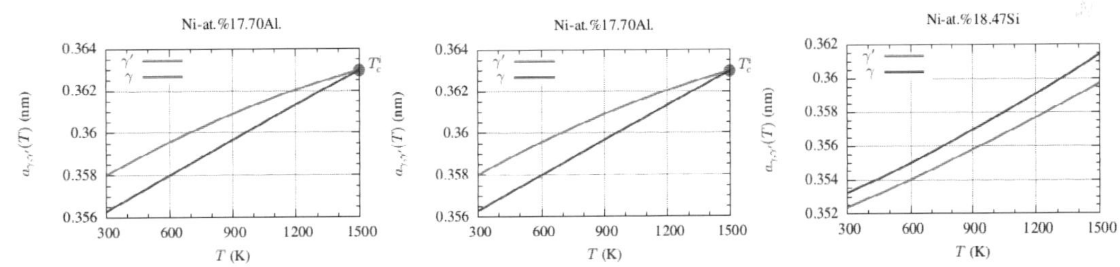

Figure 1. The temperature dependent crystal lattice parameters $a_{\gamma'}$ and a_{γ}. (a) Ni-17.70at.%Al, (b) Ni-12.40at.%Al-5.54at.%Si, (c) Ni-18.47at.%Si.

The temperature dependent misfit strain $\varepsilon_0^c(T)$ may be changed to the proximate form

$$\varepsilon_0^c(T) = \frac{2}{c_{\gamma'}(T) - c_{\gamma}(T)} \frac{a_{\gamma'}(T) - a_{\gamma}(T)}{a_{\gamma'}(T) + a_{\gamma}(T)}. \tag{6}$$

For convenience in the following discussion, the concentration of γ and γ' are assumed to be the concentration at room temperature, i.e. $c_\gamma = c_\gamma(T)_{T=300\mathrm{K}}$ and $c_{\gamma'} = c_{\gamma'}(T)|_{T=300\mathrm{K}}$. The intersecting point at $T_c = 1242.16\mathrm{K}$ in Figure 1 (b) is the critical point at which the misfit strain changes from negative to positive. Figure 2 indicates that the temperature dependent misfit strain $\varepsilon_0^c(T)$ changes along with the variation of the aging temperature T. The temperature dependent misfit strain in Figure 2 (a) satisfies

$$\varepsilon_0^c(T) > 0 \quad \left(T < T_c^i\right), \tag{7}$$

Where $T_c^i = 1499.22\mathrm{K}$ is the imaginary misfit strain collapse point. In this situation, if the temperature of starting aging T_s below the real misfit strain collapse point T_c^r, the quenching do not passes by T_c^r. Otherwise, the quenching passes by T_c^r. For the situation in Figure 2 (c),

$$\varepsilon_0^c(T) < 0 \quad T \in [T_e, T_s], \tag{8}$$

which means that the temperature dependent misfit strain is always negative during the whole quenching procedure.

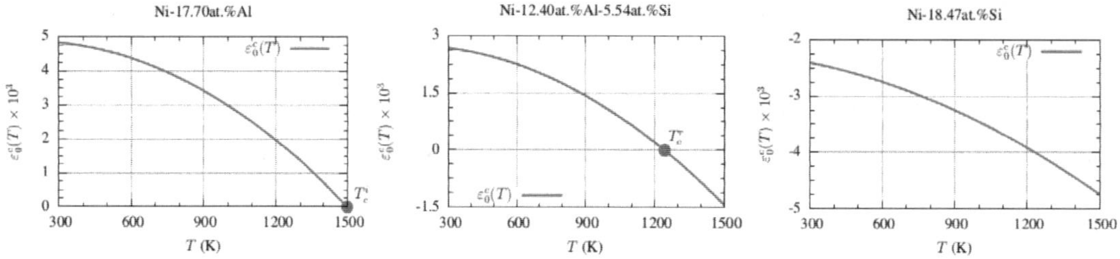

Figure 2. The temperature dependent misfit strain $\varepsilon_0^c(T)$. (a) Ni-17.70at.%Al, (b) Ni-12.40at.%Al-5.54at.%Si, (c) Ni-18.47at.%Si.

It should be noted that, the temperature dependent misfit strain $\varepsilon_0^c(T)$ demonstrated in Figure 2 (c) for the modeled Ni-18.47at.%Si alloy is always less than zero throughout the whole period of the heat treatment. The temperature dependent elastic constants for Ni-Al alloys selected from Prikhodko's work [3-4] are listed in Table 2.

Table 2. The coefficients c_i in temperature dependent elastic constant $C_{ij}(T)$ of Ni-Al-Si alloys.

	Ni-17.7at.%Al		Ni-12.4at.%Al-5.54at.%Si		Ni-18.47at.%Si	
	c_0(GPa)	c_1(GPa/K)	c_0(GPa)	c_1(GPa/K)	c_0(GPa)	c_1(GPa/K)
C_{11}	245.965	-0.0372	246.070	-0.0407	246.319	-0.0491
C_{12}	157.869	-0.0086	161.910	-0.0121	171.341	-0.0204
C_{44}	128.758	-0.0326	124.920	-0.0338	115.959	-0.0352

The coefficient error of Ni-11.17at.%Si, Ni-12.0at.%Al-5.54at.%Si and Ni-18.47at.%Si demonstrated in Table 2 is in control. The temperature dependent elastic constants of C_{11}, C_{12} and C_{44} are given by the form in Figure 3 (a), (b) and (c) respectively[3-4].

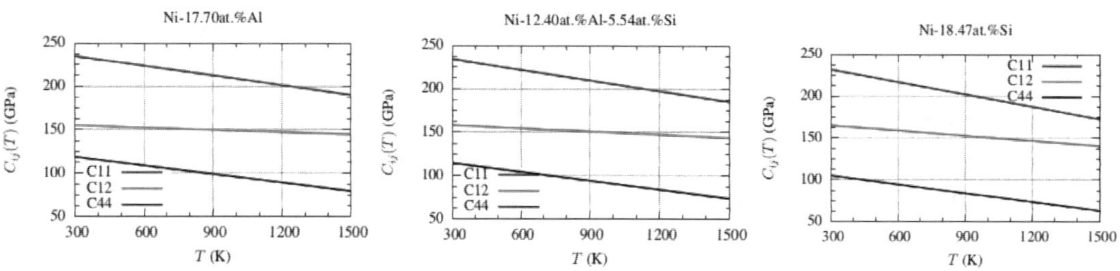

Figure 3. The temperature dependent elastic constants $C_{ijkl}(T)$. (a) Ni-17.70at.%Al, (b) Ni-12.40at.%Al-5.54at.%Si, (c) Ni-18.47at.%Si.

4. Conclusions

The MSEC is characterized by the MSEC point which is the intersection of the temperature dependent crystal lattice parameters $a_{\gamma'}(T)$ and $a_\gamma(T)$ of different phases. If the quenching passes by the MSEC point, the MSEC point is termed as the real MSEC point. Otherwise, if the quenching do not passes by the MSEC point, the MSEC point is termed as the imaginary MSEC point. Not all pairs of $a_{\gamma'}(T)$ and $a_\gamma(T)$ intersects at some point. If $a_{\gamma'}(T)$ and $a_\gamma(T)$ intersects at the imaginary MSEC point, e.g. $T_s < T_c^i$ in Ni-17.70at. %Al alloy, the misfit strain at the temperature of starting aging is not zero. If $a_\gamma(T)$ and $a_\gamma(T)$ intersects at the real MSEC point, e.g. $T_s \geq T_c^r$ in Ni-12.40at.%Al-5.54at.%Si alloy, the misfit strain at the temperature of starting aging might be approximate to zero. So that if the misfit strain is approximate to zero at T_s, the generation of the crystal nucleus need not

IOP Conf. Series: Materials Science and Engineering **761** (2020) 012015 doi:10.1088/1757-899X/761/1/012015

overcome the misfit strain energy barrier, furthermore the growth of the nucleus is only controlled by the interfacial energy, which is the reason of the dendritic morphology. The misfit strain energy may increase or decrease along with the decline of the aging temperature during quenching, which is the reason why the dendritic particles split into octet of cube.

5. Reference

[1] A. G. Khachaturyan, Theory of Structural Transformations in Solids, Wiley, New York, 1983.
[2] A. G. Khachaturyan, S. Semenovskaya, T. Tsakalakos, Elastic strain energy of inhomogeneous solids, Phys. Rev. B 52 (22) (1995) 15909-15919 (Dec. 1995).
[3] S. Prikhodko, J. Carnes, D. Isaak, A. Ardell, Elastic Constant of a Ni-12.69at.%Al Alloy From 295K to 1300K, Scripta Materialia 38 (1) (1997) 67-72 (Dec. 1997).
[4] S. V. Prikhodko, Y. Ma, A. J. Ardell, D. G. Isaak, J. D. Carnes, S. Moser, Elastic constants of f.c.c cubic and L12 Ni-Si alloys: Composition and temperature dependence, Metall and Mat. Trans. A 34 (9) (2003).

AUTHOR INDEX